Textile Science and Clothing Technology

Series Editor

Subramanian Senthilkannan Muthu, SgT Group & API
Hong Kong, Kowloon, Hong Kong

This series aims to broadly cover all the aspects related to textiles science and technology and clothing science and technology. Below are the areas fall under the aims and scope of this series, but not limited to: Production and properties of various natural and synthetic fibres; Production and properties of different yarns, fabrics and apparels; Manufacturing aspects of textiles and clothing; Modelling and Simulation aspects related to textiles and clothing; Production and properties of Nonwovens; Evaluation/testing of various properties of textiles and clothing products; Supply chain management of textiles and clothing; Aspects related to Clothing Science such as comfort; Functional aspects and evaluation of textiles; Textile biomaterials and bioengineering; Nano, micro, smart, sport and intelligent textiles; Various aspects of industrial and technical applications of textiles and clothing; Apparel manufacturing and engineering; New developments and applications pertaining to textiles and clothing materials and their manufacturing methods; Textile design aspects; Sustainable fashion and textiles; Green Textiles and Eco-Fashion; Sustainability aspects of textiles and clothing; Environmental assessments of textiles and clothing supply chain; Green Composites; Sustainable Luxury and Sustainable Consumption; Waste Management in Textiles; Sustainability Standards and Green labels; Social and Economic Sustainability of Textiles and Clothing.

Hafsa Jamshaid • Rajesh Mishra
Editors

Knitting Science, Technology, Process and Materials

A Sustainable Approach

Springer

Editors
Hafsa Jamshaid
School of Engineering and Technology
National Textile University
Faisalabad, Pakistan

Rajesh Mishra
Faculty of Engineering
Czech University of Life Sciences Prague
Prague, Czech Republic

ISSN 2197-9863 ISSN 2197-9871 (electronic)
Textile Science and Clothing Technology
ISBN 978-3-031-44926-0 ISBN 978-3-031-44927-7 (eBook)
https://doi.org/10.1007/978-3-031-44927-7

This Springer imprint is published by the registered company Springer Nature Switzerland AG
The registered company address is: Gewerbestrasse 11, 6330 Cham, Switzerland

Paper in this product is recyclable.

Contents

Introduction: Knitting Fundamentals

Hafsa Jamshaid and Rajesh Mishra

1 History of Textile

The term textile is a generic term referring to any flexible material which is produced using the network of fibers, yarns, or fabrics. Textile is a very important aspect of human's life and it plays a very important role in the human civilization. It is the basic need of the humans. The history of textile is as old as the human civilization. The manufacturing of cloths is among some of the oldest traits associated with the use and necessity of humans. Human was compelled environmentally and socially to seek out better material than leaves of trees and hides of animals to handle these factors. To protect himself from extreme environmental circumstances, he was interested to find materials which will help him in his survival. So, the history of textile was started with human civilization and textile enriched itself with the development of civilization. In sixth and seventh century BC, Excavation of Swiss lake gives the indication that the inhabitant of Swiss lake used flex and wool fiber in their clothes. Silk was introduced in India in 400 AD while spinning of cotton traces back to 3000 BC. Fabrics first appeared in the Middle East during the Stone Age, and there is evidence to support the wearing of clothes as far back as 500,000 years, although those would have consisted of skins, furs, and reeds. The earliest sewing needle to have been found dates back to France about 19,000 BC but possible needles have been found which date back an astounding 40,000 years. Dyed flax fibers found in a prehistoric cave in Georgia, Eurasia have been dated to 36,000 years ago. Evidence has been found of weaving in the Czech Republic, in the form of

H. Jamshaid (✉)
School of Engineering and Technology, National Textile University, Faisalabad, Pakistan
e-mail: hafsa@ntu.edu.pk

R. Mishra
Faculty of Engineering, Czech University of Life Sciences Prague, Prague, Czech Republic

H. Jamshaid, R. Mishra (eds.), *Knitting Science, Technology, Process and Materials*, Textile Science and Clothing Technology,
https://doi.org/10.1007/978-3-031-44927-7_1

impressions of textiles, baskets, and nets on clay dating back as far as 27,000 years. Human nature has the curiosity to have better living standard. Evolution in living standard is still continued and human is trying to have much better than this [1].

Textile fabric can be manufactured from fiber, yarn, and filaments. There are many methods of fabric manufacturing, but three of them are famous, i.e., weaving, knitting, and non-woven. Different cultures used different types of clothing depending upon social, religious, environmental fashion, and eco-systems. The trends and techniques of fabric changed according to the mentioned factors.

2 Knitting

Knitting is the second largest and most growing technique of fabric manufacturing. This technique of fabric manufacturing works by converting yarn into loops which are connected or intermeshed together to form a large fabric sheet. Knitting is a vast field and comprises two main techniques, i.e., warp knitting and weft knitting. Weft knitting is most widely used for casual wear applications. Each type of fabric product is manufactured by a specific type of knitting machine. Depending upon the end application of a knitted product, machines are categorized into hand knitting, circular knitting, socks knitting, and gloves knitting. Basically, knitting is divided into two main types. These two types are differentiated by the movement of yarn in fabric formation as shown in Fig. 1. If the yarn moves in the width of the fabric or in crosswise direction perpendicular to the length of the fabric or the direction of fabric formation, then it is known as weft knitting. If the direction of yarn propagation is in lengthwise of fabric or parallel to the direction of fabric formation, then it is known as warp knitting. Machines which are used for producing weft-knitted structures or fabrics are known as weft-knitting machines and those which are used for producing warp-knitted fabric are known as warp-knitting machines. The classification of knitting is shown in Fig. 2.

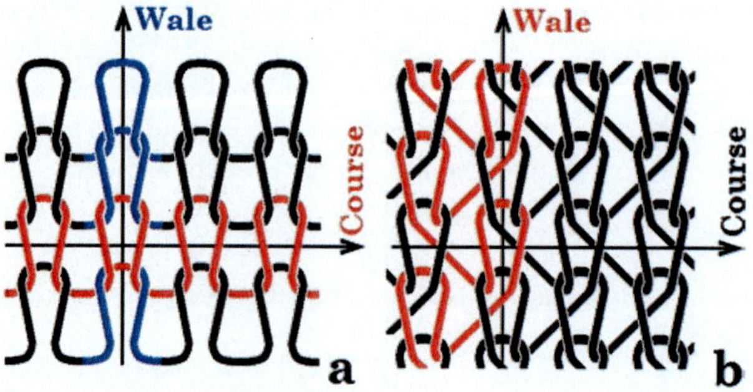

Fig. 1 Weft and warp knitting yarn propagation

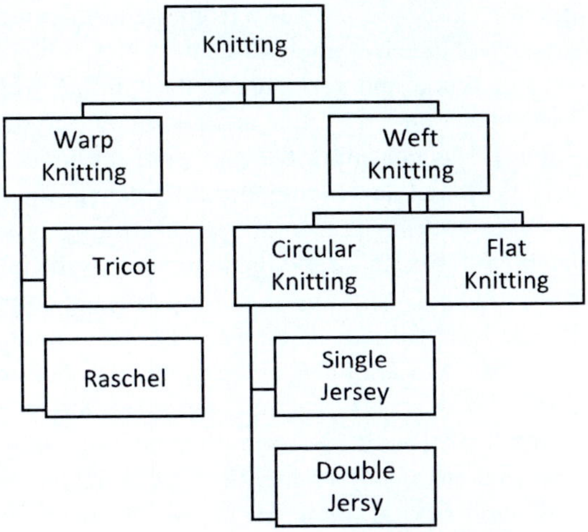

Fig. 2 Classification of knitting

2.1 *History of Knitting*

The art of hand knitting has been practiced since thousands of years. How this art was learnt by ancient human is still a mystery and so is the country and time of its origin. However, some believe that this art originated in Persia. Others claim Israel, Jordan, and Syria belt as its origin, and still others claim mountains of North Africa. It is claimed that the earliest known knitted items were produced by Muslims who were employed by Spanish Christian Royal families. They were highly experts in knitting different types of items. Their knitted items are found in the tomb in the Abbey of Santa María la Real de Las Huelgas in Spain. Knitted socks discovered in Egyptian tombs have been dated between the third and sixth century BC.

In ancient times, knitting was done with the help of wooden sticks which were known as knitting needles. These knitting needles were used by the person to make loops of yarns. This was known as hand knitting. Hand knitting is the oldest method of producing knitted fabric because machines were invented much later. The basic tool used for loops of yarn was two needles or pins [2].

A detailed first history book of hand knitting was written by Richard Rutt in 1987 [3] named "A History of Hand Knitting." According to the writer, tubular stocking stitching was the first form of knitting and purl stitch was the first stitch that was invented in earlier than 1562 and was used in turning the heel stocking. Later on, after the mid-sixteenth century and onward, purl stitch was used for decorative purposes. Early knitting was done in round form but still unclear the number of needles used.

Flat knitting was the second knitting technique used to convert yarn into fabric. The knitted items of Mediaeval Egyptians were flat but flat knitting of Europeans was probably influenced by framework knitting. Flat knitting is unique in a sense that garments can be made on it directly. It is likely that knitters were imitating clothes tailored by garments manufacturing. There are examples of fine jackets imitating brocade probably from the 17th century. The fine jackets were imitated brocades since seventeenth century. By nineteenth century, flat knitting became very common and garments of flat knitting were seen everywhere. The shapes of garments were very crude and complexity in design was introduced till the 1930s. After 1930s charming designs were introduced in knitting and designer became very skilled in developing elegant practical shaping [4].

Warp knitting is also a type of knitting in which knitting yarn runs along the length of fabric in a zig-zag pattern to knit the fabric. Tricot and Raschel are the two basic types of warp knitting. Warp knitting comprises many types of technical and leisure wear fabrics. The basic difference between warp and weft knitting is that adjacent wales are made at the same time instead of single row or course. Since the number of strands of yarn are converted into loops simultaneously, it is not possible to make warp-knitted fabric by hand as is done in weft knitting. Warp knitting is always done by a machine. In 1589, William Lee, a clergyman from England, who invented stocking frame. He applied for patent in the same year for making knitted article and laid the foundation of mechanical manufacturing of warp-knitting machine. Development of warp frame was made by Josiah Crane in 1775. The first warp-knitting frame was sold by Josiah Crane in 1778 to Mr. Richard March who patented it. These early machines were modifications of the stocking frame with an additional warp beam. These warp frames were used for any kind of thread and transverse thread can be inserted by an optional element. Dawson got the credit for inventing cams that move the bars and which help to regulate the twist in 1796. Brown and Copstake were able to succeed in imitating Mechelen net. In 1799, bobbin was invented by Lindley. Robert Brown patented the first twist frame of warp knitting that was able to knit a wide net fabric [5].

In 1947, an intuitive entrepreneur and mechanic, Karl Mayer introduced the first warp-knitting machine. The FM 48 was comprised of two guide bars, and it used bearded needles and attained the maximum speed of 200 rpm. It was marked as the beginning of a technical era in ground-breaking jumps in the field of warp knitting.

Karl Mayer in 1953 introduced the first Raschel warp-knitting machine and presented for sale in the market. This warp knitting machine consisted of four guide bars, four take-up roller and a pattern box with a pattern chain and disc. In 1954, the first elastic raschel and the first tulle raschel warp-knitting machines were introduced in the market. In 1955 "Super Garant" series was introduced in the market and in 1956, first Lace warp knitting machine with 12 guide bars was introduced. Later more new development was introduced in the market. 1958 first curtain Raschel and in 1959 first carpet Raschel warps knitting machines were presented in the market. The series of inventions didn't stop and in 1967 the first fall plate multiguide bar raschel machine was introduced in the market. Later on, jacquard raschel machine with weft insertion facility and many other features was presented in the market. Following this series of inventions, the production range diversified and the lace raschel machines' quality was improved. The lace raschel machine was able to work with 57 guide bars in 1981. The electronic control system was also added in the same era. In the beginning of 1980s, lace warp-knitting machine was loaded with an electromechanical pattern guide bar control named summator [6, 7].

3 General Terms of Knitting

Basic terms of knitting which are associated to the structure are explained in this section. These terms are widely used and easy to understand. Without the understanding of these terms, analysis of knitted structure and any knitted fabric is not possible.

- *Knitting Loop or Knit Stitch*

Loop is a basic unit of knitting. It is formed by the knitting needle and is a curved shape of yarn. Two main parts, needle and sinker, are used for making a loop in fabric. Therefore, loop is divided into two parts. The upper portion of the head and side legs are considered as needle loop while the sinker loop is that part which connects the loop to adjoining loops also known as the foot of the loop as shown in Fig. 3.

- *Face Loop and Back Loop*

When a loop is formed, its process of manufacturing w.r.t. reviewer tells that it would be a face loop or back loop. When a new loop forms, if it passes from the back side to the font of the previous loop (towards the viewer) it is known as face loop. Face loops only show their legs on the right side of the fabric. When a new loop passes to the back from front of the previous loop during interloping, then it is known as back or reverse loop. Back loop shows only the head and foot portion of the loop/stitch. The loop structures are shown in Fig. 4.

- *Machine Pitch and Machine Gauge*

Machine pitch is related to the gaps between needles in the needle bed of machine. It is mostly measured in needle gaps per inch. Machine gauge is almost a similar term used for defining fineness of the machine. Machine gauge is elaborated as no. of needles per inch in any needle bed of the machine. Machine gauge is usually used for defining the range of yarn counts which could be used on that machine.

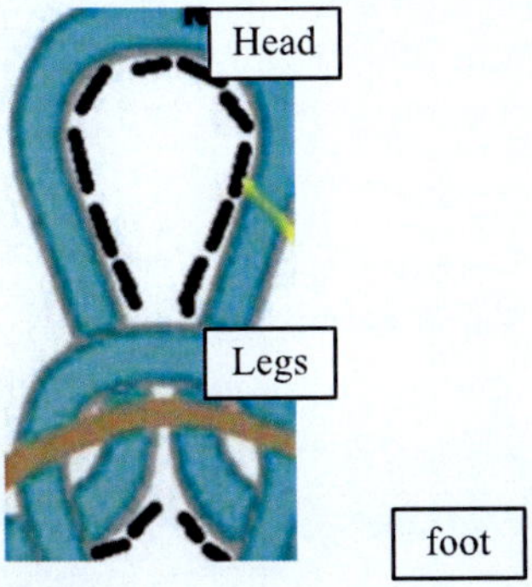

Fig. 3 Parts of knit loop

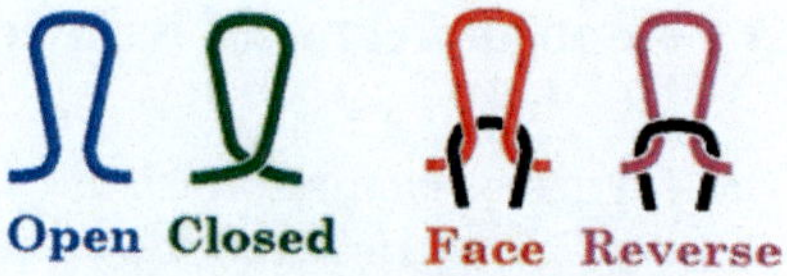

Fig. 4 Face and back loop

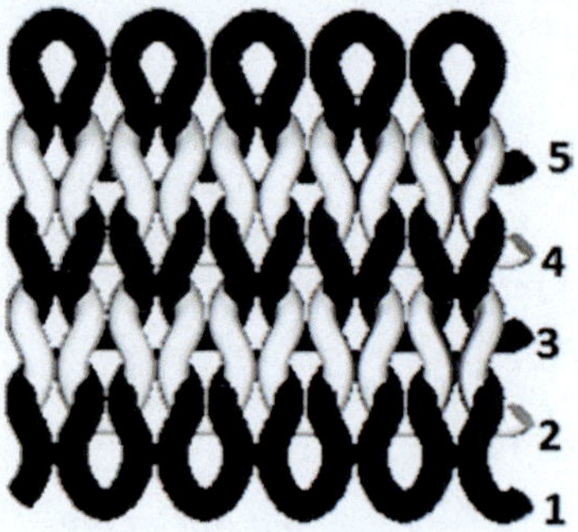

Fig. 5 Identification of courses

Machine pitch = No. of gaps between needles/inch
Machine gauge = No. of needles/inch

- *Courses*

Course is a row of yarn loops inserted by all needles widthwise in the fabric in one revolution of the machine. Same as one pick/weft is inserted in the woven fabric. Courses are also measured in per inch and per centimeter area. It is known as C.P.I. (courses per inch) and C.P.cm (courses per centimeter).

No. of courses fed in one revolution of knitting machine would be equal to the no. of feeders of machine. In Fig. 5, the courses are shown.

- *Wales*

Wales are vertical columns of yarn loops in the length direction of the fabric as shown in Fig. 6. All loops present in one wale are manufactured by one needle in successive knitting cycles. Wales are equivalent to the warp yarns inserted in a woven fabric. No. of wales present in the width of fabric are equal to no. of needles operating in a machine. Wales are measured as wales per inch or per centimeter. Denoted as W.P.I. and W.P.cm.

- *Stitch Density*

Stitch density is the no. of stitches present in the defined area. Mostly it is measured in per inch square or centimeter square. Stitch density is calculated by the multiplication of no. of wales per inch and no. of courses per inch. In Fig. 7, the stitch density is 25.

- *GSM*

GSM is the weight of fabric per square meter.

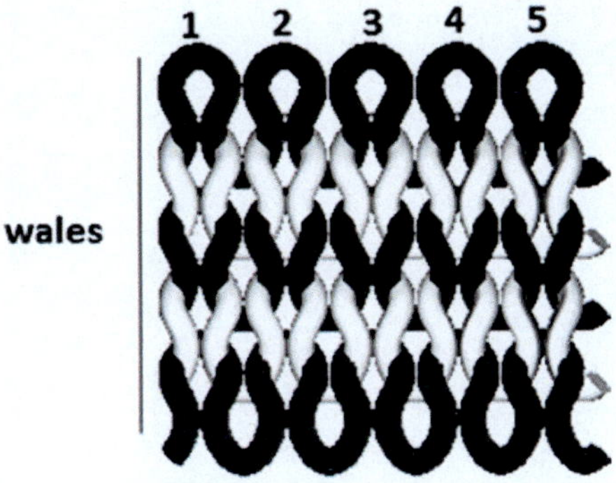

Fig. 6 Identification of wales

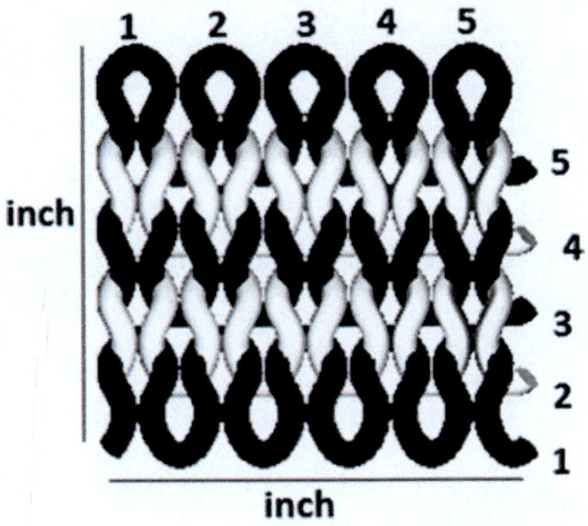

Fig. 7 Stitch density

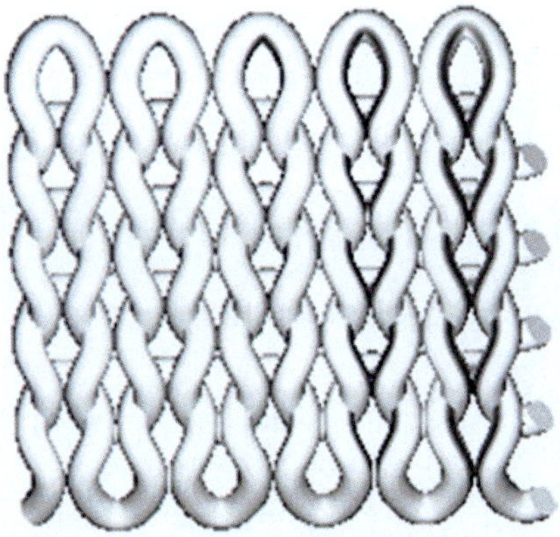

Fig. 8 Technical face side

- *Technical Face and Technical Back*

Technical face and technically upright side of knitted fabric is that side which comprises of legs parts of the knit loops, which appear as "V" shape, in other terms side which contains all face loops.

Technical back side is that side which contains all back loops and where head and foot appear on the fabric. Both types of structures are shown in Figs. 8 and 9, respectively.

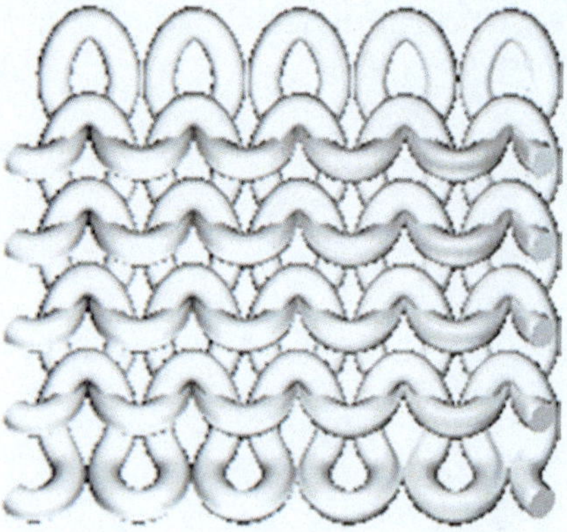

Fig. 9 Technical back side

- *Stitch Length*

Stitch length is that length of yarn which is consumed by one loop in its formation. Stitch length is the basic parameter upon which properties of the knitted fabric depends. All properties of fabric can be engineered by the adjustment of stitch length. It is measured in centimeters and millimeters. Stitch length is denoted by S.L.

4 Comparison of Weft and Warp Knitting

Weft knitting and warp knitting both are completely different fabric manufacturing technique although both techniques intermesh loops of yarn. In both these techniques, knitting machines, preparation of yarn for knitting, and working of machine for fabric formation are different. Warp-knitted fabric is also different from weft-knitted fabric in the aspect of properties of fabrics. General comparison of weft- and warp-knitted fabrics is given in Table 1.

5 Merits of Knitting

- Minimum number of yarns required to knit the fabric; even single yarn can be used to make a fabric.
- Properties of knitted fabric such as air permeability, fabric areal weight, shrinkage, thermal resistance, extensibility, etc. can be engineered by adjusting the size of the loop.
- By adjusting the loop size, fabric porosity and dimensional stability can also be managed.

Table 1 Comparison of weft and warp knitting

Sr. no.	Weft-knitted fabrics	Warp-knitted fabrics
1	Only one yarn can be used for making fabric and also maximum no. of yarn depends upon no. of feeders of machine	Large no. of yarns are required to produce fabric as every needle takes an individual yarn
2	Loop/stitch formation occurs in the width direction of the fabric	Loop formation occurs lengthwise in the warp-knit fabrics
3	Every loop in horizontal direction is made of the same yarn	Every loop in horizontal direction is made of different yarn
4	Yarn is supplied in the form of cones or cheese	Yarn is fed in the form of yarn beam same as weaving known as warp beam
5	There is no additional process after spinning	Warp beam preparation includes an extra step
6	Spun yarns are mostly used as raw material for weft knitting	Filament yarns are used in warp knitting according to the required end use. Also, anti-static oiling is required here due to static charges occurrence in synthetic yarns
7	Weft-knitted fabrics are less dimensionally stable	Warp-knitted fabrics are dimensionally more stable
8	Higher elasticity	Less elastic fabric
9	There is less variety of making different designs in machines	Wider variety of different designs can be made in these machines
10	Mostly latch needles are used	Latch, bearded, and compound needles are used
11	Machinery can be circular or open width both	It has only flat or open-width machines
12	Weft-knitted fabrics are more suitable for the apparel purpose	Warp-knitted fabrics are more suitable for technical and strength purpose

- Large number of designs can be made by adjusting the position of cams and needles.
- Different face and back structures, different face and back colors can be made on knitted fabric.
- Large number of conventional and technical fabrics can made through knitting.
- Knitted fabric is much comfortable due to easy movement of loops and is mainly used for leisure wear.
- Wastage of yarn is minimum, and as a result, cost of production is low.
- Moisture can be well managed by using plating technique in knitting, and this fabric is used for summer wear especially for sports persons.
- Using extensible yarns, fabric fitting is adjustable with body size. So, same fabric can be used for different body size.

5.1 Demerits of Knitting

- Knitted fabric has less dimensional stability and its dimensions can be changed with time and external conditions.
- Fabric unravels quite easily even if single loop is damaged.
- Knitted fabric looses its shape after stretching and cannot fully recover to its original shape.
- High-performance and brittle yarns such as glass, Kevlar, and Nomax are very difficult to knit because of their high bending rigidity.
- Fabric curled from the edges especially from the single knit fabric.
- Due to low twist in yarn, knitted fabric is prone to pilling.
- The raveling edge of knitted fabric creates difficulties in sewing process.
- Low count range for the same guage machine. To change the yarn count, we have to change the guage of machine.

6 End Uses of Weft and Warp Knitting

Application areas of weft-knitted fabrics and warp-knitted fabrics are different. Weft-knitted fabrics are softer, flimsy, and thick fabrics in comparison with warp-knitted fabrics. Weft-knitted fabrics are mainly used in clothing. Weft-knitted fabrics are mostly used in the apparel of daily life due to their softness and comfort properties. They are not as such recommended for technical applications due to their lower dimensional stability. Main products of weft-knitted fabrics are tee shirts, polo shirts, trousers, sweaters, knitted scarves, undergarments, socks, gloves, tablecloths, and mats. Weft-knitted fabrics are sometimes used in some technical applications also. Weft-knitted fabrics are used as upholstery, knitted blankets, medical stockings, and bandages.

It has already been discussed that warp-knitted fabrics are higher in strength and dimensional stability. Due to these characteristics, they are preferably used in technical applications while a small portion of these fabrics is used for esthetic and functionality in clothing industry too. While warp-knitted fabrics are used in apparel in the manufacturing of fashion articles and as an accessory (laces, fancy fabrics). Laces, fine fabrics for undergarments, safety vests, cloth lining, and shoe linings are made of warp knitting. Warp-knitted fabrics are mostly used in technical textiles due to greater strength, dimensional stability, and their fineness.

Warp knitting is also used in the manufacturing of medical products like knitted hernia mesh, artificial heart valve, and artificial blood vessels. In the construction industry, warp-knitted fabric is used as reinforcements for pavements, river banks, concrete, and for the reinforcement of soil which are known as geotextiles. Agriculture and horticulture also used knitted fabrics which are in the form of nets used to save plants from extreme weather conditions, birds, and for tying the hay bales.

References

1. Rudi, Volti, *The Facts On File Encyclopedia of Science, Technology, and Society, E-N. Facts on File, Volume 2* (New York Publishers, 1999)
2. S.C. Ray, *Fundamentals and Advances in Knitting Technology*, 2nd edn. (Wood Head Publishing Limited, 2012)
3. R. Rutt, *A History of Hand Knitting*, 1st edn. (Anchore Press Limited, 1987)
4. Lyer, Mammel, Schach, *Circular Knitting*, 3rd edn. (Meisenbach, Bamberg, 2004)
5. S. Raz, *Flat Knitting Technology*, 1st edn. (Universal Maschinenfabrik, 1993)
6. D.F. Paling, *Warp Knitting Technology*, 2nd edn. (Columbine Press Ltd, 1965)
7. P. Dawson, *Encyclopedia of Knitting Techniques*, 1st edn. (Addison-Wesley, London, 1984)

Yarns in Knitting

Awais Ahmed Khan, Hafsa Jamshaid, and Rajesh Mishra

1 Introduction

Yarn is a continuous strand of staple fibers or filaments twisted together to keep them intact, offering strength and flexibility simultaneously so that it remains suitable for further fabric manufacturing process. Yarn is the first and foremost element in knitting to produce knitted fabric of any type. Different yarns are composed of different types of fibers which are joined together with the help of different types and degrees of twist which are imparted with the help of various spinning methods. Different yarn manufacturing system such as ring, rotor, air jet, etc. impart different structures and properties to yarn. The standard definition of yarn according to the international standard ASTM D 4849 is "A generic term for a continuous strand of textile fibers, filament or material in a form suitable for knitting, weaving or otherwise intertwining to form a textile fabric." Normally, there is no requirement for any outer binding material in the formation of yarns and keep their structure intact as the fibers their self are adhered together by various diverse techniques. This is the reason that yarns are flexible in nature and provide a wide range of diverse techniques for yarn manufacturing. In the present era, requirements of consumers are increasing with respect to properties. The most important of these are comfort, appearance, and its hygiene features. So, the textile product should be eye catching by designs or color, soft and smooth appearance. Products should have thermo-physiological

A. A. Khan (✉)
Faculty of Textile Engineering, National Textile University, Faisalabad, Pakistan

H. Jamshaid
School of Engineering and Technology, National Textile University, Faisalabad, Pakistan

R. Mishra
Faculty of Engineering, Czech University of Life Sciences Prague, Prague, Czech Republic

H. Jamshaid, R. Mishra (eds.), *Knitting Science, Technology, Process and Materials*, Textile Science and Clothing Technology,
https://doi.org/10.1007/978-3-031-44927-7_2

comfort besides ergonomic comfort. So, at present, researchers are trying to develop different fibers or improving the existing fibers for desired properties.

2 Selection Criteria for Weft Knitting w.r.t. Fibers and Yarns

Properties of different inputs and outputs during the manufacturing process are greatly interlinked and play their part in deciding the properties of final products as illustrated in Fig. 1. So, to achieve the better-quality knits, it is very crucial to select the optimum fibers and yarns as their processing has a huge impact on the knitted fabrics.

Fiber is the building block of yarn, and yarn is the basic unit of knitting. Prior to the knitting, it is very necessary to choose the yarn from a wide range of fibers. Fiber's properties are very important that play a vital role in the performance of knitted fabrics. The properties of final products are greatly influenced by the raw material. A broad range of fibers can be used in the development of knitted products having different properties and textures such as softness, elasticity, stiffness, roughness, smoothness, etc. Yarns which are appropriate for knitting should have enough strength, resiliency, elasticity, and moisture properties. These properties which make yarn suitable for knitting mainly depend on the type of fiber and construction of yarn. A wide collection of natural fibers can be easily used in knitting such as wool, cotton, silk, flax, etc. The development of synthetic manmade fibers also made enough room to be utilized for the knitting products. Among synthetic fibers, polyester, acrylic, nylon, and rayon fibers are most commonly used in knitted fabrics.

Commonly used fibers in knitting can be categorized into three major categories:

- Animal fibers: wool, silk, angora, mohair, alpaca
- Plant fibers: cotton, flax, ramie, linen, bamboo
- Manmade fibers: polyester, nylon, rayon, acrylic

Animal Fibers Wool fiber is commonly used for knitting in animal fiber category as wool offers good properties to be used in knitwear. Wool is commonly used in

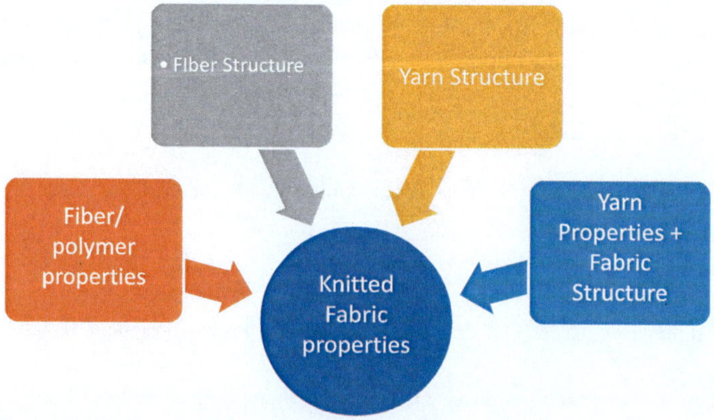

Fig. 1 Parameters affecting the knitted fabric properties

winter knitwear like sweaters, jerseys, outerwear, socks, gloves, etc. However, there is little difficulty while knitting wool because of weakness in the wool fibers against various parts of knitting zone. So, yarn strength should be ensured to avoid excessive breakages which may lead to holes in the fabric structure. In this context, fiber crimp is very important to be considered to provide better elasticity and strength. To avoid such problems, it is better to use wool fibers blended with some other fiber.

Plant Fiber Among plant fibers, the cotton fibers in knitting can be found since 1730. In knitting industry, the most popular fiber is cotton which can be used on almost all kinds of machines. It can be used for almost all types of apparels. Cotton-knitted products are suitable for summer wear. Cotton is a natural fiber offering good moisture and comfort properties and is readily utilized in knitted innerwear and underwear.

Synthetic/Manmade Fibers In the list of synthetic fibers, polyacrylonitrile (Acrylic), polyamides (Nylon), and polyester are commonly used. *Acrylic* fibers are the most commonly used manmade fibers in knitting industry especially for garments and socks. The regular use of acrylic fibers is owing to their strength, better resilience properties, lightweight, better handle, and softness characteristics. Acrylic fibers also offer a better blending tendency toward the wide range of natural fibers and can be used readily as an alternative to wool fiber. However, acrylic fibers lack wicking properties which somehow offer limitations in their use. Acrylic fibers are often used to produce texture in the knitwear which cannot be offered by natural fibers. Nylon is among some popular synthetic fibers which are being used in knitting. Knitted fabrics made of nylon offer good strength, elasticity, abrasion resistance, and lightweight. Nylon fibers are used in blends with wool. They have also applications in hosiery items, sportwear, and underwear. Polyester is also one of the common fibers used for all types of knitted garments [1].

Some important aspects of fiber which are necessary to be considered and lay the foundation of selection criteria for knitting are discussed in the present section. Properties of fibers should be considered before selecting the fibers as they affect the fabric properties. Physical properties of fibers characterize the physical behavior of fiber. Physical properties which are very important to be consider are: dimensional characteristics (length, cross sectional, diameter of fiber), moisture sorption (moisture content , regain, absorption desorption, adsorption), linear density, friction (it is an important aspect which form the basis dependent on surface characteristics and crimp), optical properties (tendency of fiber to reflect light), thermal properties, and *mechanical properties* (tensile properties, elastic recovery, relaxation, and creep behavior) [2].

The fibers used in pure form or blended form behave differently. Some commonly used fibers for knitted articles are cotton, polyester, wool, etc. as mentioned before. Yarns having 100% pure content of one fiber type are homogeneous yarns, while in blended yarns different percentages of fibers are blended together to achieve the desired properties which could not be possible otherwise. A consequential phenomenon in case of blended yarn is fiber migration which plays a very crucial role in deciding the properties of yarn and end-use product. The final properties of yarn will be greatly affected by the positioning of fiber. For example, in commonly used blended yarn of cotton and polyester, if the polyester fibers migrate toward the center of yarn, it will result in the presence of mostly cotton fibers around the surface of yarn. So, the yarn will apparently behave as cotton yarn in terms of surface characteristics such as better comfort and moisture absorption, less abrasion, etc. as the polyester being positioned at the center will provide sufficient strength.

Reciprocally if the mostly cotton fibers migrates toward the center and polyester being positioned along the surface, in this case the resultant yarn and fabrics will behave differently. Yarn will have different feel, handle, and luster due to polyester being positioned on the surface. The final knitted fabric will have more likely to higher pilling rate.

3 Classification of Yarns

Knitting machine efficiency mainly depends on yarn quality, knitting parameters, and human skills. So, it is very crucial to select the optimum fibers and yarn types to achieve the better efficiency and quality of knitting. Textile yarns are diverse in their nature, properties, manufacturing techniques, and their uses. Many authors have categorized yarns in different appropriate ways. Here we classify yarns into main two categories for better and easy understanding, and then further subcategories will come under these major categories as shown in Fig. 2.

- *Staple spun yarn*
- *Filament yarns*

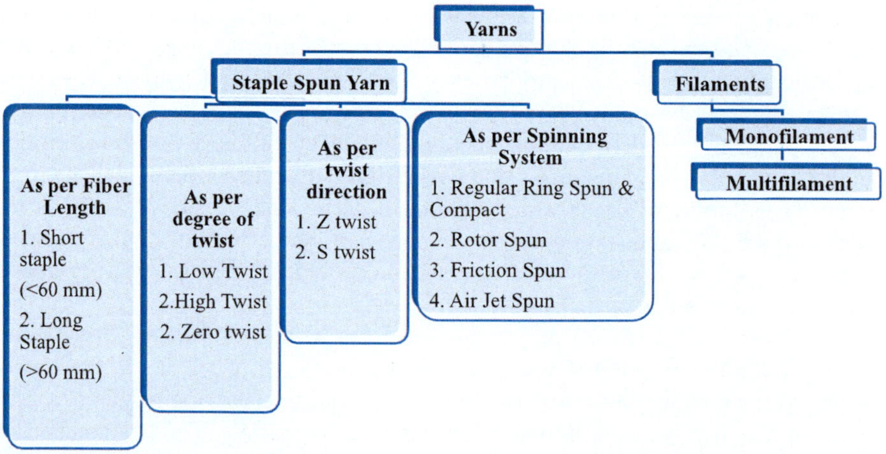

Fig. 2 Classification of yarns

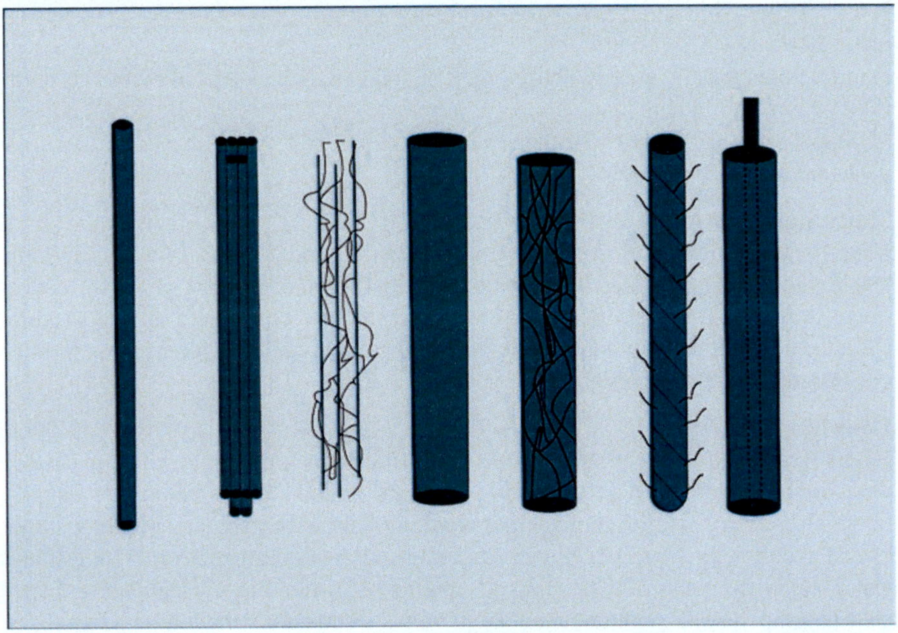

Fig. 3 Physical appearance of different yarns: (**a**) Monofilament, (**b**) Multifilament, (**c**) Comingled yarn, (**d**) Tape yarn, (**e**) Fibrillated tape yarn, (**f**) Spun yarn, (**g**) Core yarn

Physical appearance, which is shown Fig. 3, can be helpful for identifying different yarn types.

3.1 Staple Spun Yarns

Staple yarns are normally natural fibers and produced by consolidation/integration of staple fibers. Staple spun yarns are composed of definite length of fibers held together with the help of twist or any other mean. Fibers are adhered together in yarn body by means of different techniques or methods. In staple spun yarns, fibers are held together by the mechanical forces applied by the twist or wrapping. The fibers in spun yarns have distinct nature, so when they are held together, they still have variable compactness which results in air pockets in the spun yarns. These air pockets perform better insulation when used in the knitted fabrics for outer wear or winter wear and also give better comfort and warmth feeling. Spun yarns have lower lustrous, uneven appearance, hairy surface, and more absorbent.

Spun yarns may be made by natural or blends of natural and synthetic fibers depending on requirements.

Staple spun yarns are further divided into subcategories based on staple length, degree of twist, twist direction, and manufacturing technique as mentioned in Fig. 2.

3.1.1 Types of Staple Yarn Based on Fiber Length

Natural fibers usually exist in short length or staple length except silk which is naturally occurring filament.

Staple spun yarn can be further categorized into two types, that is *short staple yarn and long staple yarns.*

Short Staple Yarn

Usually, staple fibers are in the ranges between 10 and 500 mm. In case of short staple yarn, the staple fibers have the maximum length of 60 mm, while for cotton fibers the length of fibers between 25 and 45 mm are considered as short staple. Cellulosic fibers usually come under the category of short staple as they usually occur naturally in short length.

Long Staple Yarn

Fibers with length higher than 60 mm are considered as long staple, while in case of wool long staple length is about 60–150 mm. Animal fibers which are usually obtained from the hair of animals like wool, mohair, cashmere, etc. are long staple fibers, and the yarns thus developed from these fibers are categorized as long staple yarns. Apart from naturally occurring staple fibers, some filaments are cut into specific length to use as staple fibers in spun yarns. In this case, the filament extruding from spinneret are initially formed into tow comprising hundreds of filaments, and then these are texturized and cut into specific staple length. These filaments are texturized while this process to improve the fiber adhesion which is necessary during spinning.

3.1.2 Types of Staple Yarns with Respect to Twist Amount/Degree of Twist

In spun yarn, twist is a critical factor which tends to keep the fiber adhered together in the yarn. Optimum twist level is very critical for the better strength of yarn. As the strength of yarn increases by increasing the twist up to a certain level, then above this point again the strength of yarn starts decreasing. In case of knitting, usually low twist yarns are recommended because high twist become the main reason of various knitting faults. The amount/degree of twist in yarn can be adjudicate by assessing the number of times the fibers are twisted together per unit length. The degree of twist in a yarn depends on the diameter of yarn. The amount of twist very much below the recommended standard result in fiber slippage in yarn. Much low twist in yarn may cause more pilling. Zero twist yarns are most favorable for knitting to produce the articles in which softness is the top priority. Lord et al. studied open-end, twist less, and ring yarns with 100% cotton and polyester cotton blends and made single jersey fabrics and assessed their performances for spirality. They found that open-end yarn has maximum spirality and zero twisted yarns have minimum or zero spirality. The reason for that is residual torque which is responsible for spirality [3, 4]. Less twist is used for the production of loose fabrics such as bath towels, baby blankets, and various plush fabrics.

3.1.3 Types of Staple Yarns with Respect to Twist Direction

Apart from the degree of twist, direction of twist has its own impact on the yarn properties. Yarns can be classified on the basis of twist direction. Staple yarns are further classified into three types based on the direction of twist. The direction of leaning line of fibers tells whether the yarn has S twist or Z twist as shown in Fig. 4.

S-Twist Yarn

The fibers in yarn are twisted together in clockwise *direction* and fibers leaning line form the inclination shape of English Alphabet S then it is called S Twist.

Z-Twist Yarn

The anti-clockwise or counterclockwise twist in yarn is termed as Z twist. The inclination of fibers in the yarns seems like the English Alphabet Z.

The direction of twist has a prominent impact on the spirality behavior of knitted fabrics. As researchers found that the curling behavior of single jersey weft-knitted fabrics produced using Z-twist yarns is higher than the fabrics knitted from S-Twist yarns [5]. It was also found in literature that the spirality in the knitted fabrics is affected by the direction of the twist of yarn [6].

The twist direction in the yarns has a very prominent effect on the spirality of knitted fabrics. Z-twist yarns tend to bend the wales toward right making Z skew. Alternatively, S twist in the yarns result in the left side skewing of the wales. It is also revealed by the researchers that the twist direction in the yarns also has a solid relationship with the rotation of knitting machine. So, to reduce the spirality, Z twist yarns should be used on the machine rotating counterclockwise and S twist yarns should be used on the knitting machines rotating in clockwise direction [7].

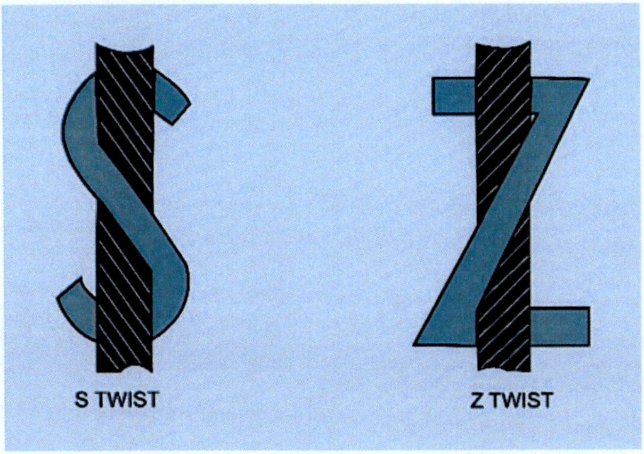

Fig. 4 Yarn twist identification

3.1.4 Types of Staple Yarns w.r.t. Manufacturing Technique

On the basis of spinning technology, spun yarns are further classified into different categories which are explained in the next sections. Manufacturing techniques also affect the properties of knitted fabric. It was also found that knitted fabrics composed of combed cotton yarns have lower thermal resistance values as compared to knitted fabrics composed of carded yarns [8].

The two spinning methods that dominate the industry are ring spinning and open-end/rotor spinning (ring-spun yarn and rotor-spun yarn are shown in Fig. 5). In case of surface properties such as yarn evenness and hairiness, conventional combed yarns have better surface properties as compared to carded compact yarns with greater evenness and low hairiness [9].

Ring-Spun Yarn

Ring-spun yarn is known for the commonness in spun yarn technology. Ring spinning is the most versatile technology in the history of yarn manufacturing. Ring spinning is the most widely acknowledged yarn manufacturing technique which was first developed in the 1830s in America. Ring spinning has a diverse nature in its technology as it can produce yarns with a great variety of counts and twists from a wide range of fibers. This is a continuous spinning technology in which a tiny circulating traveler is used to insert twist while winding is also done simultaneously. Ring spun yarns are high in strength and have low imperfections along with better/higher bursting strength. Ring yarns show good performance for knitting as it is found that ring spun yarns show highest strength among all the spun yarns. Ring spun yarns are the most diversified yarns which can be used in a wide range of knitted products: from underwear, T-shirts to sweaters for outer summer wears.

Covered yarns or core-spun yarns (Fig. 6) are mostly used in knitting to impart the desired characteristics to knitted fabrics. Use of different materials in core and Sheath in core-spun yarns led to their diverse applications. Use of different materials in core and sheath in core-spun yarns led to their diverse applications. Use of elastomeric filament in core covered with cotton fibers (core sheath yarn is illustrated in Fig. 6) are used to produce stretch fit knitted garments. Core-spun yarns can be manufactured using various spinning techniques including friction, rotor, and air jet.

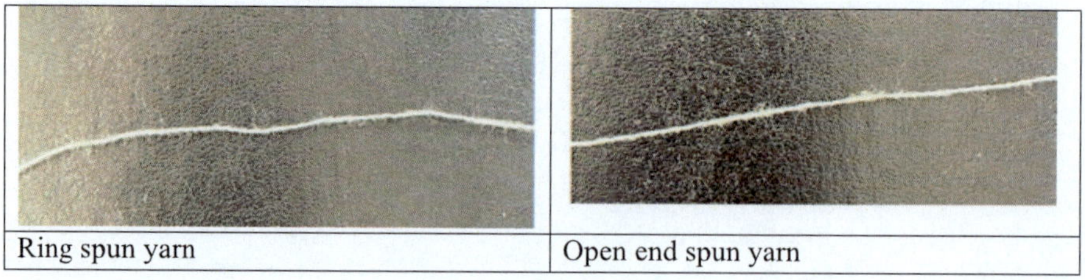

Fig. 5 Type of yarn

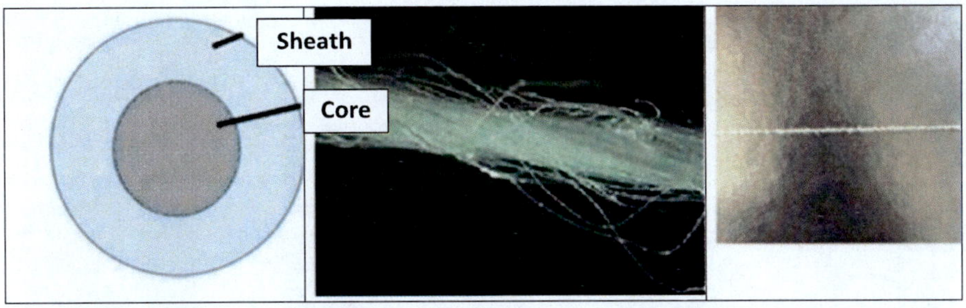

Fig. 6 Cross section of core-spun yarn

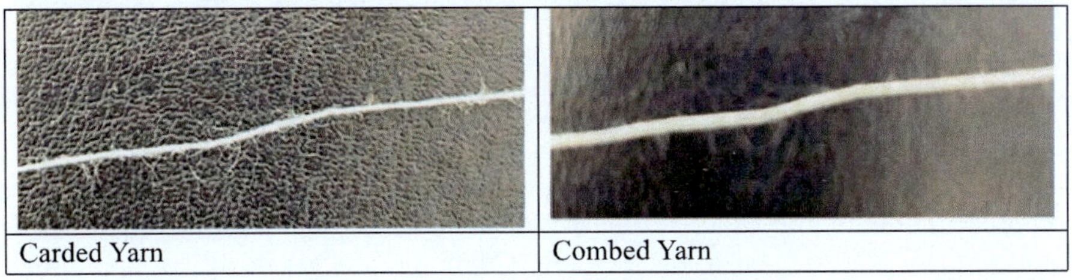

Fig. 7 Ring-spun yarn

Core yarns are generally composed of two components: the core which is inner part and sheath which is outer. Core-spun yarn consists of at least two materials. Cover yarn with filament or spun is placed in core and covered by staple fibers. Cover yarn having higher strength with better comfort properties can be produced by using high strength filament in core and natural fibers used for covering. Ring spinning provides a good opportunity for the manufacturing of cover yarns used for technical purposes. The ring-spinning frame can be equipped with core-spinning attachment. These yarns are also used in technical products; for example, cut-resistant knitted gloves are produced using steel wire as a filament in core and staple Kevlar fibers used as cover. Use of elastomeric filament in core covered with cotton fibers is used to produce stretch fit knitted garments. Core-spun yarns can be manufactured using various spinning techniques including friction, rotor, and air jet. Other technical yarns used for fire resistant or electrical properties can be produced using ring-spinning system.

Ring spun can be further categorized into combed and carded yarns (carded and combed yarns are depicted in Fig. 7) based on the process involved during their manufacturing and their properties.

Carded Ring-Spun Yarns

Carded yarns are typical single yarns characterized by the higher short fiber content; also carded yarns lack high fiber orientation. Carded yarns do not undergo the additional combing process. Carded yarn is very commonly used yarn. Carded yarns have low cost and economic yarns for a wide range of applications.

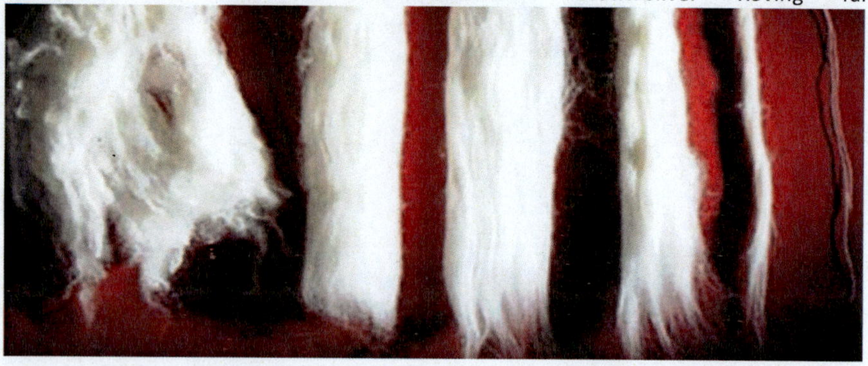

Fig. 8 Flow process of yarn

Combed Ring-Spun Yarns

In addition to the carding process, ring-spun yarn is subjected to additional and useful process of combing which results in achieving the more better yarn quality (different stages of combed yarn; Fig. 8). Combing process involves the further alignment of fibers and removal of short protruding fibers. Combing process further eliminates 5–25% of short fiber content as compared to carded yarns. As mentioned, combed yarns have more fiber alignment which tends to increase the yarn strength and elongation at break and reduces the number of faults and imperfections in the yarn. Combed yarns are of high quality with better surface characteristics and more absorption and softness. Combed yarns are used to produce high-quality knitted fabrics. Knitted fabrics produced using combed yarns have a very better pilling resistance. Selection of combed yarn for knitting is done by evaluating the combed yarn on the basis of noil percentage. Noil percentage is the percentage of waste removal during combing process. Noil parentage ranges from 4% to 24%. Greater the removal of noil percentage result in higher quality of yarn. It also determines whether the yarn is fully combed or partially combed. Fully combed yarn is composed of all combed slivers while partially combed yarns are composed of carded and combed sliver. Less twist is used to combine the long-combed fibers in combed yarn which will be beneficial in preventing the spirality effect in knitted fabrics.

Compact Ring-Spun Yarns

Yarns produced using conventional ring-spinning system have high hairiness which influences the quality of yarns. Compact spinning is a variation/modification of ring spinning that condenses the roving before the final twist which leads to improved smoothness and strength. It is also called condensed spinning. In conventional ring spinning, the formation of a spinning triangle is a common problem. To overcome this problem, compact ring spinning was innovated. It reduces the size of the spinning triangle to a minimum. The hairiness of yarn is thus reduced, yarn evenness is improved, and tenacity is higher when compared to ring-spun yarns.

Rotor-Spun Yarn/Open-End Yarns

Rotor spinning is the second most widely accepted technique of yarn production after ring spinning. This method eliminates the formation of roving. In rotor spinning, the (rotor: a high-speed centrifuge) rollers and air streams are used to separate fibers strands from sliver feed, and then again, these fibers are recollected in a rotor grove. The fibers are distributed around the rotor circumference and held there for a while by centrifugal force. The yarn is withdrawn from rotor wall and twist is produced due to rotation. Rotor-spun yarns are composed of short fibers as compared to the ring-spun yarns. Rotor-spun yarns are bulkier in nature or heavy yarn and having lower hairiness as compared to ring-spun yarns. Rotor-spun yarns are relatively weaker, so they tend to cause more and frequent yarn breakages during knitting. Rotor spun yarn has a stiffer structure, that's why it shows poor drape properties. This yarn is more preferred for toweling, denim, and pile fabrics

Friction Spinning

It is a method of open-end spinning which uses the two rotating rollers of which at least one of must be perforated so that air can be drawn through it surface to facilitate fiber collection. High yarn rotational speed is achieved by the friction between the roller surface and the yarns.

Air Jet-Spun Yarn

Nowadays, a new spinning system, *Jet-Ring-Spun Yarn* has been developed to overcome the issues encounter in ring spinning. In this system, an air jet nozzle is placed under the conventional ring-spinning system to reduce the hairiness of conventional ring-spun yarn; this system is called jet-ring-spinning system. It is found that jet ring yarns have low hairiness values as compared to the conventional ring-spun yarns. These jet ring-spun yarns have higher properties than conventional ring and compact yarns in terms of hairiness, tensile properties, etc. [10].

The yarns produced through pneumatic method of spinning are called air jet-spun yarns. In this method of yarn manufacturing, air jet flow is used to impart twist in the fibers. Among all spun yarn manufacturing techniques, air jet spinning is the fastest technology. It is possible to produce yarns with more controlled properties but of more complex structure. There are two layers in the structure of air jet yarns, the inner core comprising parallel fibers and covered by the wrapping fibers around.

Air jet yarns are further subdivided into two types:

- Conventional air jet spinning/Murata jet spinning
- Murata vortex spinning

Conventional Air jet-spun yarn: In Air jet-spun yarns, long staple fibers are preferred as there is no true twist in yarn rather than parallel fibers in core are held with the air of wrapper fibers. The long staple fibers provide better holding and wrapping of fibers in the structure. Air jet-spun yarns are characterized by the false twist. False twist is inserted with the help of two nozzle air jet spinning. The yarn structure resembles the fascinated yarn. Conventional air jet spinning has its own limitations regarding use of fibers for the manufacturing of yarn. It preferably uses 100% man-made fibers and blends of cotton and synthetic fibers. Purely cotton fibers cannot be used as they don't have enough strength to withstand the yarn manufacturing processes. It is found that knitted fabrics produced using air jet yarns have high pilling resistance.

Vortex-Spun Yarns: Vortex spinning technology was introduced by the Japanese company Murata Ltd that's why it is also known as Murata vortex spinning. It is different from air (false twist) spinning. Vortex spinning is the advanced version of two nozzle jet spinning and this does not imply of false twist. A rotating air vortex twists the free fiber ends around the middle/bridge fibers with true twist, producing a ring yarn type of structure. Vortex spun yarns have lower imperfection index and hairiness, lower elongation, and higher tenacity because of regular and comparatively closer packing of fibers. Vortex yarns have higher pilling resistance. The knitted products made of vortex yarns resulted in less shrinkage. In vortex-spun yarns, 15–30% twisted surface fibers are present. In terms of drape-ability of vortex yarn, knitted fabrics were poorer than ring and rotor yarn-knitted fabrics [11].

3.2 Filament Yarn

Filaments yarns are man-made fiber except silk and produce by extrusions methods. Continuous filaments of individual length are held together by means of twist or without twist to form filament yarn. Filament yarns are comparatively smoother, uniform, lustrous, and more hydrophobic in nature. The greater crystalline nature of filaments result in lower moisture absorption and better uniformity. Filament yarns are comparatively more durable. Filament yarns are mostly prepared from the thermoplastic polymer while it is also possible to produce filament yarn from natural origin such as cellulose. It is the matter of extreme concern that cellulose cannot be melted to be used for further spinning, so it is mixed with compatible solution for further processing. Rayon is an example of filament yarn produced from cellulosic origin. Filament yarns are also used as reinforcement in knitted fabrics. Such as in double knit fabrics or spacer fabrics where, commonly polyester filament yarn is used for tucking between two layers of the fabrics.

Filament yarns are composed of one or more filaments of indefinite length throughout the yarn body. Filament yarns are further categorized into two major categories based on the number of filaments present in the yarn body, i.e., monofilament and multifilament.

Monofilament

The yarn composed of only one single filament running throughout the length is called monofilament yarn as shown in Fig. 9. These yarns are considered simplest form of yarns available for fabric manufacturing. These monofilament yarns are mostly solid and circular in cross section. Although monofilament yarns with modified cross section can be produced by changing the shape of spinneret hole as per

Fig. 9 Monofilament yarn

requirement. Various types of monofilament yarns are used in knitting to gain the desired properties. Elastomeric spandex (Lycra ®) (shown in Fig. 9) is an example of most commonly used elastomeric monofilament yarn in knitting to impart extra stretch properties. Other than spandex, nylon and polyester monofilaments are also used commonly in knitting. Monofilament yarns are also used to produce technical knitted products, such as fishing nets, medical textiles, and industrial workwears. Their application as technical yarns are common like steel monofilament yarn (monofilament steel yarn shown in Fig. 9) [12].

3.2.1 Multifilament Yarn

Multifilament yarns can be further classified into two broader categories.

Flat Yarns/Tape Yarn

Filament yarns obtained from the extruder are in smooth and flat form. Multifilament yarns having low level of twist are known as flat yarns. Flat yarns have higher smoothness and luster, low stretch, lower bulk, and lower covering power. Often texture is imparted in flat yarns for their use in apparel. False twisting and untwisting are done to impart texture and require crimp, elasticity, and stretch. Flat yarns offer a very limited range of applications.

Texture Yarns

Filament yarns are mostly flat and smooth throughout their length. Filament yarns also don't have enough stretch due to higher crystalline nature. Elastomers are exception in this case as they are more stretchable. In order to be properly utilized in various end applications, texture is added to the filament yarn after extrusion process. So, in this context, texture yarns are defined as "a multifilament yarn which produced through extrusion spinning process is subjected to further processing to impart coils, crimps, loops and other distortions along the length of filament which result in increased bulk and stretch in the nature". Some commonly known techniques are used to add texture into the yarns. Such as knife edge, air jet, stuffer box, false twist, and knit de knit [13].

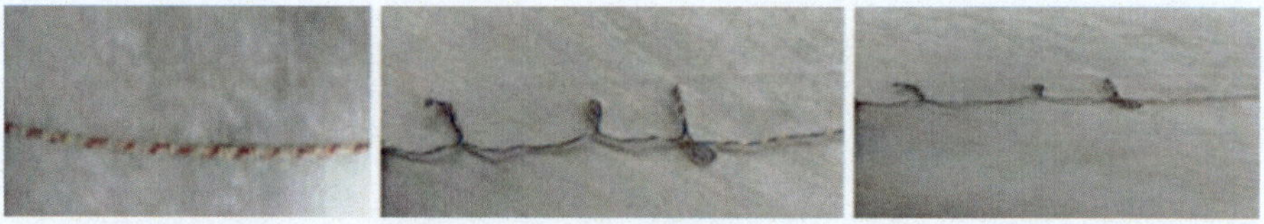

Fig. 10 Marl yarn and loop yarn

3.3 Fancy Yarn

Fancy yarns impart special textures, decorative, and aesthetic features to the end products. Fancy yarns can be made of natural or synthetic fibers, either staple or filament. There are various types of yarns having their own characteristics. Few of them are mentioned in this section and depicted in Fig. 10.

Marl yarn: Marl yarn is composed of two different colored yarns twisted together during doubling process. This doubling gives a fancy combination of two different colors at alternative intervals. These yarns are usually utilized to produce knitted fabrics with simple structure and irregular color patterns and subtle effect. This is the most simple fancy yarn.

Corkscrew yarn: Corkscrew yarn is actually a plied yarn which is composed of two yarns. Among these two yarns, one yarn spirals around the other result in giving the fancy texture. It is also called spiral yarn.

Gimp yarn: Gimp yarn is also bi-component yarn which comprises of a twisted core yarn and surface yarns. The outer yarn which is wrapped around the core yarns gives the rippled effect. It is produced in two stages: first twisting both yarns around each other and then reverse twisting. For better stability, binder yarn is also required.

Diamond yarn: Diamond yarn is composed of multicolored filaments folded around a coarser single yarn or roving. In this yarn, one filament is folded around the coarser yarn using S-twist while the other with Z-twist which ultimately give a colored pattern in diamond look. There is expression of compression (due to filament) on the coarser yarn as well. Diamond yarns are very useful for producing knitted fabrics with multicolored patterns using simple structures.

Snarl yarn: Snarl yarn is composed of core and effect yarn, in which high twisted yarn is wrapped around the core and exhibits twist projection from the surface of yarn. So, because of this effect, the high twisted yarn is overfed which results in snarls-like texture.

Loop yarn: Loop yarn is a multi-component yarn composed of four yarns in which two yarns made up the core, one effect yarn and one binder yarn to keep the structure stable. All the yarns should be of optimum quality and requirement. Cores consist of two-ply yarn around which the effect yarn is wrapped. Effect yarn is overfed to give the effect of loop like projection on the surface of yarn. The loop size depends on various parameters such as overfeed rate, spinning tension, twist level, and the space between the drafting rollers. Sometime slivers are also used as the core for the production of loop yarns.

Knot or spot yarn: Knot yarns are known from different names such as nub, spot, or knot yarns. These yarns are characterized by tiny, raised segments along the length. These tiny, elevated sections are produced by the addition of extra yarn or fibers twisted in a particular way. These yarns are used in the production of knitted hats, sweaters, and blankets to impart a decorative texture into the knitted product [14].

4 Quality Requirements of Yarn in Knitting

By focusing on improving the quality of fabrics, doing things right at first time also mean reduction of process which leads to positive effect on environment. As it will affect reduction in rejection which help in minimizing the use of stripping agents to remove color, improving the dye-ability of the fabric. Also, optimum use of energy and other resources leads to better utilization of resources. Durable knitted textile products demand higher quality in all aspects. So, the durability and quality of knitted products get influenced by very important parameters which must be considered and ensured. The visual illustration of the quality of end product and the factors influencing them are given in Fig. 11.

Effect of yarn parameters on the properties of knitted fabrics.

Yarns are the important part of knitted products (raw materials) which has very strong effect on the knitting during the process and later performance of knitted products. Yarn quality affects fabric quality, performance, and cost. Selection of an

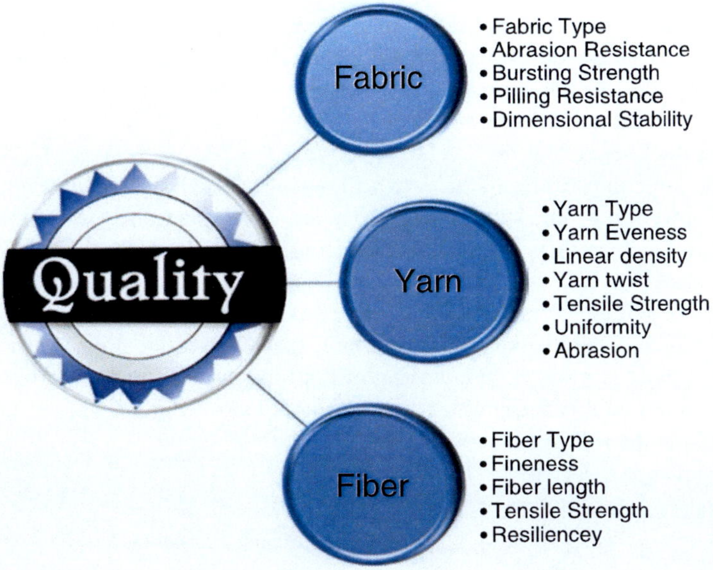

Fig. 11 Knitted fabric quality parameters

ideal yarn that match the requirement of product is a great deal to make prior to the knitting. Yarn quality refers to whether the yarn meets the minimum requirements of the knitter or not. Yarn quality requirement is changing every day. It is easy to make the highest quality yarn just for the sake of achieving the best yarn results. But it is difficult to produce a good-quality yarn with a minimum deviation. Very high fluctuation in yarn quality is an evil for any end use. Sometimes it is better to keep the same level of yarn quality (around 25% USTER STANDARDS) by strict quality control than achieving 5% USTER STANDARD but without consistency. Consistent quality is an important for knitted fabric quality.

During the knitting process, yarn has to pass through various stresses and strains like twisting, friction, and bending. So, for the achievement of good quality knitted fabric, it is very necessary to ensure the yarn quality parameters. Normally, there is no problem regarding the variation in the quality of filaments. However, variation in the quality of spun yarns is a big problem. To produce quality and cost-effective knitted fabrics, it is mandatory to use the yarns of optimum or desired quality. The use of yarns which do not meet the required quality standards leads to the more wastage results in lowering the efficiency and increasing the rejection.

Parameters of yarn decide the yarn properties, so ultimately these parameters influence the knitted fabric properties. Various yarn characteristics have prominent impact on the behavior of knitted fabrics. For example, knitted fabrics are prone to structural distortion which is mainly due to the residual torque in yarn. This residual torque in yarn is due to twist multiple (TM), fiber type, and yarn count. Twist coefficient is an important parameter of yarn. Yarns used for knitting usually have low twist having low value of TM as compared to the TM required for weaving because these yarns do not need to be as strong as weaving yarns and therefore need no sizing and less twist. This lower twist leads to softer yarn and fabric. Yarn torque or liveliness should be at a minimum to help prevent excessive fabric shrinkage, skew, and torque. The researchers found that higher values of yarn twist coefficient result in lowering the thermal resistance as when the yarn twist coefficient increases the yarn become finer and result in lowering the thickness of fabric which ultimately lower the thickness of fabric.

Good elongation values in the yarn will reduce fabric holes. Good evenness values will prevent machine stops and fabric holes. Thick places in the yarn need to be minimized because they can lead to yarn tension problems, broken needles, and bent latches [15].

There are some factors that are majorly responsible for the proper knittability of the yarns such as uniformity in yarn count, proper yarn cleaning, yarn strength, amount of twist in yarn, proper waxing of knitting yarn is an essential factor. Vital factors of yarn quality requirements for development of quality knitted fabrics are mentioned in the following subsection.

4.1 *Yarn Count/Numbering System*

Globally acknowledged yarn numbering system is very much necessary for the consolidation of the yarn size measurement. To determine the yarn thickness of different types and to evaluate the yarn, a specific number is assigned to the specified thickness keeping in view the length and weight ratio. This specific number is called count of yarn. Some related terms such as yarn number, yarn count, yarn size, and yarn linear density are used for the same purpose. Conclusively, yarn count shows the fineness or coarseness of yarn. The ultimate purpose of yarn numbering system is the determination of its linear density. Textile materials are usually sold in weight, so it is very necessary to express the yarn weight with respect to its length and thickness. Generally, there are two ways of saying this, how much weight of a given length of yarn? Or what length of the yarn in given weight?. So widely acknowledged, there are two systems which describe the yarn linear density directly or indirectly and they are named as direct system and indirect system. Different yarn count number systems use specific units for weight and linear measurement and expressions. Use of units appropriate for the respective yarn numbering system is of prime importance as the measurement of yarns involves very large figures related to length or weight. Otherwise, the measurements will become overwhelmingly large or too small. So, use of specified units suggested by scientists is very important and necessary. Both yarn numbering have their respective advantages and disadvantages. In the direct system, yarn number is called the linear density of yarn with units of Tex, Grex, and Denier. While in the indirect system, yarn number is called the yarn count with units of Nec, Nm, and N woolen. Both yarn numbering systems are inverse of each other. So, in this section the yarn numbering will be explained according to this categorization.

4.1.1 Direct System

In direct yarn numbering system, yarn number and size are in direct relation with each other. In direct count system, yarn weight is evaluated per unit length of the yarn.

$$\text{Direct yarn number} = \text{Weight} / \text{Length}\,(\text{Constant})$$

Yarn length is given in meters and weight in grams. So, it means length is constant and weight is variable; as a result, the count will increase as the weight/thickness of the yarn increases. Commonly known direct yarn counts are Denier, Tex, and decitex. Filament yarn number is usually expressed under the direct numbering system

Denier
Denier is commonly used for synthetic fibers, filament yarns, and silk. This unit length of 9000 m was suggested by technologists.

$$\text{Denier} = \text{No. of grams} / 9000\,\text{m}$$

Here one Denier is the weight of standard length of 9000 m of yarn. So, 45 denier yarn means 45 g weight of yarn for every 9000 m length of yarn. The yarn size or coarseness increases as the Denier increases, so it is very easy to understand that yarn of 90 denier is about twice the coarser than yarn of 45 denier. It is matter of very logical and easy interpretation that in the cross section of 90 denier staple fiber yarn on average there are 30 fibers of 3 denier approximately.

Tex
In this subsystem of direct yarn count, the weight of yarn in grams of 1000 m length of yarn is described as 1 Tex. In this system, the yarn becomes coarser as the Tex increases.

$$\text{Tex} = \text{No. of grams} / 1000\,\text{m}$$

Grex
American society of testing materials (ASTM) proposed and appreciated the system of direct yarn count called Grex. Grex is the weight of yarn in per 10,000 m of yarn.

$$\text{Grex} = \text{No. of grams} / 10{,}000\,\text{m}$$

4.1.2 Indirect Numbering System

Indirect yarn numbering system is a fixed weight system. It is a major consideration that in indirect system yarn is measured in terms of length per unit weight. In the indirect system, yarn number and yarn thickness are inversely proportional. This means that as the yarn count increases, the yarn thickness decreases and hence yarn becomes finer. The name of indirect system is based on this indirect relationship.

$$\text{Indirect yarn number} = \text{Length} / \text{Weight}\ (\text{Constant})$$

So special units should be used as mostly there are various systems which have been bestowed from the ancient crafts and commonly there are no concrete logic behind the selection of these units. Within these two broadly classified systems of yarn count, there are some subsystems in these categories as well.

Normally yarn counts are defined as hanks/lb in this count system while each subsystem has different specified length of yarn.

English Cotton system	Nec = No. of hanks/lbs
Worsted system	$N_{worsted}$ = No. of hanks/lbs
Woolen system	N_{woolen} = No. of hanks/lbs

For example, in cotton count system and those systems which evolved from cotton processing technology, hank is a specific length comprising 840 yards. For common understanding, if the count of cotton single yarn is 30, it is written as 30s or 30/1, in other words 20 hanks/lb. So, it will consist of 20 × 840 yards in one pound of yarn. Nec symbol is used to differentiate the cotton count from others and is also known as *English count*. In *wool count* system, hank has a specified length of 256 yards; similarly for clarification symbol, N_{woolen} is used for woolen yarns. *Worsted Count* ($N_{worsted}$) is also an indirect yarn count system which is expressed as the number of hanks per pound of yarn. Where one hank is the length of 560 yards of yarn. In this system, the fineness of yarn increases as the yarn count increases.

Metric count is represented as Nm which is meter/grams. Length is 1000 m.

$$\text{Metric system Nm} = \text{No. of hanks} / \text{Kg}$$

The growing business of spinning results in development of various yarns to describe and evaluate the yarns. For example, there are Galashiels Cut, Yorkshire Skein woolen (YSW). Apart from these, for carpets yarns Dewsburry Counts are also used. So different yarns used different units for measurement and different terms for various descriptions. For example, in Yorkshire skein system, the length of one skein is 256 yards and Galshiel Skein length of each skein is 300 yards for linen. These are further explained in Tables 1 and 2. There are some conversion constants for yarn count conversion as shown in Table 3.

Yarn weight choice varies with respect to the final product. Heavy yarns are usually used to knit fabrics used for winters. Thickness and weight also play critical part in the finishing of the final article. Single yarns are widely used in knitting, but plied or double yarns offer wide range of products with better properties. Use of plied yarn in knitting eliminates the spirality problem in knitted fabrics balancing the torque produced by the twist.

4.1.3 Yarn Count for Plied Yarns

Plied yarns are composed of two or more yarns twisted together. Plied yarns have compound yarn count system. Showing both yarn count and number of plies. Plied yarns are commonly used in knitting. In direct count system, 100 denier/3 means that 3 plies of 100 denier yarns were twisted together to give a yarn. The density of this plied yarn will be equal to 300 denier as in direct count system the increase in number of yarn count depicts the increase in linear density yarn. In direct count system, linear densities of individual yarns are simply added together to get the total linear density of plied yarn. On the other hand, in indirect count system, the count

Table 1 Yarn standard measuring length in different indirect count systems

Count system	Standard weight	Standard length
English Cotton	1 pound	Hank of 840 yards
French Cotton	1 Kg	Hank of 1000 m
Bump Cotton	1 oz	1 yard
Spun Rayon	1 pound	Hank of 840 yards
Spun Silk	1 Pound	840 yards
Worsted	1 pound	560 yards
Woolen (Yorkshire skien)	1 pound	256 yards
Woolen- American Cut	1 pound	Cut of 300 yards
Woolen-Dewsburry	1 oz	1 yard
Woolen-Galashiel	24 oz	Cut of 300 yards
Woolen-West England	1 oz	Snap of 320 yards
Woolen – Alloa	1 pound	Spyndle of 480 yards
Linen (wet spun)	1 pound	Lea of 300 yards
Hemp, ramie	1 pound	300 yards
Asbestos – British	1 Pound	Hank of 50 yards
Asbestos – American	1 Pound	Cut of 100 yards
Glass fiber		100 yards

Slater (1945), "Indirect System of Yarn Numbering and Calculation – Textile Calculations" (n.d.)

Table 2 Equivalent yarn specifications

Metric (NM)	Yorkshire skien weight	Worsted	Yards per ounce	Galashiels cut
1	1.875	0.875	31	2.5
3	5.75	2.75	94	7.5
5	9.75	4.5	155	12.5
7	13.5	6	215	17
9	17.5	8	280	22
11	21.5	10	345	27.5

Pl (n.d.)

Table 3 Yarn count conversion table

	Direct count		Indirect count	
	To denier	To tex	To metric count	To worsted count
From denier		0.111 × denier	9000/denier	7972/denier
From tex	9 × Tex		1000/Tex	885.8/Tex
From metric count (Nm)	9000/Nm	1000/Nm		0.8858 × Nm
From cotton count (Nec)	5135/Ne	590.5/Nec	1.693 × Nec	1.5 × Nec
From worsted count (Nw)	885.8/Nw	7972/Nw	1.129 × Nw	

Wang X [16]

system for plied yarns is somehow complicated. The twist effect is ignored, and the number of yarns placed side by side in the plied yarns are imagined to be of same length of each. For example, the weights of individual yarns in plied yarns are W_1,W_2, W_3, etc., then the total weight will be $W_T = W_1 + W_2 + W_3$, etc., but in indirect system it will be calculated as illustrated in the example below:

$$W1 = \frac{L}{840Ne1}, \quad W2 = \frac{L}{840Ne2}$$

$$WT = \frac{L}{840}\{1/Ne1 + 1/Ne2 + 1/Ne3 + \text{etc.}\}$$

Here N_{e1}, N_{e2}, N_{e3}, are the yarn counts of individual yarns which are plied together. So the equivalent count of plied yarn in this case will be

$$NeT = \frac{L}{\mathbf{840}WT}$$

and

$$\frac{1}{NeT} = \frac{1}{Ne1} + \frac{1}{Ne2} + \frac{1}{Ne3} + \text{etc.}$$

For example, if there are four yarns having yarn count of 40s of each yarn are plied together and ignore the twist effect

$$\frac{1}{NeT} = \frac{1}{40} + \frac{1}{40} + \frac{1}{40} + \frac{1}{40}$$

Here, it can be seen that the compound yarn count

Yarn Numbering System and Machine Gauge Relationship

The gauge (E) of knitting machine is fixed distance between the needles of knitting machine. So specified yarn count can be used for specific knitting machine. Yarn count of various yarns is related closely with some important parameters of knitting such as machine type, needle of various types, hook size of needle, and distance between dial and cylinder. The formula to find suitability of Tex count with the machine gauge is given as

$$\text{Yarn Tex} = (100/E)^2$$

The suitability of cotton yarn count for circular knitting machine can be calculated using the equation below

$$\text{Cotton Count} = (\text{gauge})^2 / 18$$

Table 4 Relationship of structure and yarn count (experimental + knitting calculations)

Fabric type	Yarn count relation with GSM
Single Jersey	Yarn count = (−0.141) × (GSM) + 50.22
Interlock	Yarn count = (−0.206) × (GSM) + 80.56
Pique	Yarn count = (−0.146) × (GSM) + 57.16
1x1 Rib	Yarn count = (−0.123) × (GSM) + 54.57
Double Lacoste	Yarn count = (−0.167) × (GSM) + 64.36
1x1 Rib Lycra	Yarn count = (−0.119) × GSM + 59.12
2x2 Rib Lycra	Yarn count = (−0.108) × (GSM) + 56.62

Fabric areal density or GSM (gm/sq meters) is very important factor which plays a vital role in performance properties of the knitted fabrics and deciding its end use. Yarn count has very close and prominent relation with the areal density of knitted fabrics. Knitters drive various formulas/relations or equations to choose the optimum yarn count for the required areal density or grams per square meter (GSM) of the knitted fabrics. There are different fabric structures in knitting which have their own range of areal density, so the proper yarn count is recommended for the specific structure to achieve the desire GSM. In Table 4 are some relations given which are developed based on research and experimentation.

4.2 Yarn Twist

Twist is very important characteristic of yarn as optimum twist is necessary for the efficient performance of yarn. Yarn twist is the number of turns present in a unit length of yarn. Units used for yarn twist are turns per inch (TPI) or turns per meter (TPM). In knitting usually low twist yarn is preferred to use because high twist yarn will cause the knitted fabric defect known as skewness or spirality. During knitting, yarn has to passed through multiple knitting machine elements like guides, tensioners etc., so high twist in yarn will result in disturbing the alignment of wales to diagonal position which is called skewness or spirality [17].

Knit structures are formed by bending the yarn into a loop and then interlacing them to create a fabric. The curvature of loop would be smooth and well defined if the bulkiness of the yarn is higher. The bulkiness eliminates sharp bending and improves resiliency of the structure, and these fabrics are expected to stretch easily and recover during use. Thus, the actual purpose of using low twist yarn is to achieve this smooth curvature to loops and high resiliency to fabric.

Very low twist in yarn is also not acceptable as it will lead to the low strength and ultimately yarn breakage during knitting process. Twist has a very prominent effect on the different properties of yarns such as strength, handle, moisture absorption, elongation, durability, appearance, pilling property, and many more. Twist has a critical influence on the spirality in knitted fabrics. A very moderate effect of twist was found on bursting strength of knitted fabrics. There must be evenness in the twist of the yarn for the better processing of yarn during knitting, otherwise it will disturb the proper processing. Twist has also a prominent effect on the handle and softness of the fabrics. Yarns with low twist produce soft and comfortable knitted fabrics. Despite of this, yarns with low twist has efficient knit-ability because of having low stiffness [18].

4.3 Yarn Evenness/Uniformity

Yarn evenness determines the level of uniformity in the yarn linear density throughout its length. Apparently, filament yarns are smooth and uniform throughout their length, so the evenness of staple spun yarns is of major importance/concern. The lower the yarn evenness, the more frequent or percentage of presence of thick and thin places along the yarn. The higher unevenness value in yarn also leads to irregularities in the yarn count as well. Yarn unevenness has its own impact on the twist level of the yarn; this may lead toward the twist variation which ultimately affect the knit-ability and quality of end product. Uniformity in the yarn surface is very important as it also influences the strength of yarn. Thick and thin places in yarns encounter the different knitting parts during knitting process which lead to the various faults ultimately resulting in lowering the yarn strength. For obtaining smooth curvature of loop and its uniformity, the yarn should be uniform in thickness and imperfections should be minimum. The thin place in yarn receives more twist resulting in compact structure, and thus sharp bends in loop while thick place receives less twist and forms a large curvature at loop. The co-efficient of friction at thin places might be higher due to increased twist, which might influence the wax pick-up. Yarn uniformity or evenness is measured by using Uster Tester. Fig. 12 shows the appearance of neps, thick, and thin places in yarn and knit fabrics.

4.4 Yarn Hairiness

Yarn hairiness can be defined as the amount of free, protruding fibers including free fibers or fiber ends from the yarn body to outer. Spun yarns are produced from short staple fiber, so they are hairy. It means that fiber ends come out of the main body of the yarn and its surface becomes hairy. It is a type of yarn fault. It imparts fuzzy appearance to the yarn and reduces luster of the yarn. Hairiness is the quantitative parameter related closely with the diameter of yarn. Yarn hairiness is very important parameter defining the comfort and aesthetics of the final products. Yarn hairiness also effects the fluff generation on the knitting machine during knitting which effects the knitting performance. Yarns having more hairiness result in higher fluff production. The knitting performance is badly affected by its deposition at different

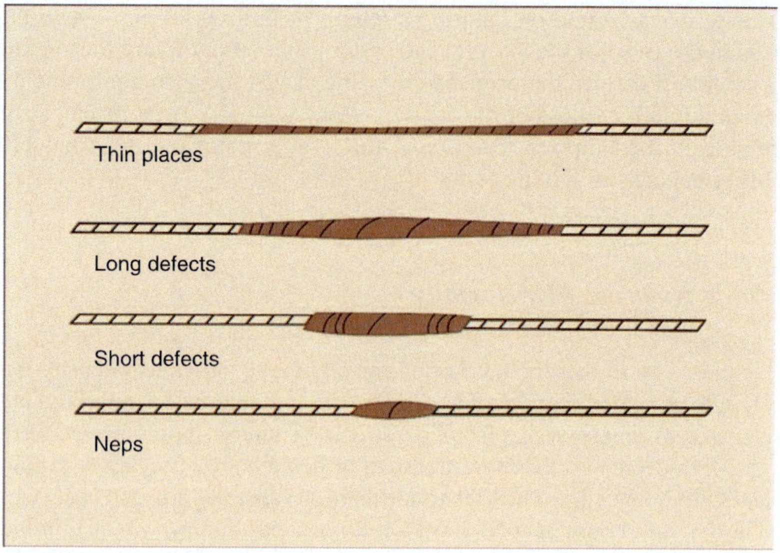

Fig. 12 Appearance of faults in yarns

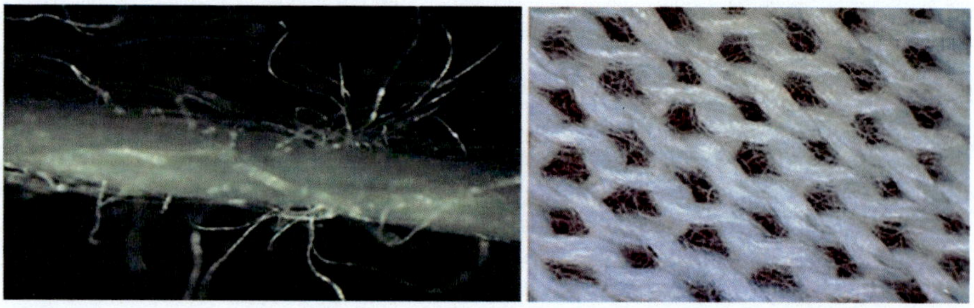

Fig. 13 Hairiness on yarn and knitted fabric

places and responsible to produce poor-quality fabric with fuzzy appearance. Yarn hairiness is also influenced greatly by the fiber distribution or mean fiber position. So if the yarn has high value of mean fiber position so it will have more fiber distribution near the surface of yarn which will tends to increase the hairiness value of yarns [19]. Hairiness is also measured by using Uster Tester which consists of hairiness module. Microscopic view of yarn hairiness is depicted in Fig. 13.

4.5 Strength

Tensile strength of yarn is one of the prime factors which determines the knit-ability of any yarn for any specific end use or application. For staple yarn, strength and elongation are measured. Strength can be measured as either single end yarn strength or count leas strength product (CLSP).

Single end strength method is more likely used in the industry nowadays due to its reliability and authenticity. Tensile testing of yarns is used to determine the breaking force, elongation, and breaking tenacity. Breaking tenacity, a ratio of the breaking force to yarn linear density, is used for comparison and validation purposes. It is equivalent to Rupture in Km (RKM). Uster Tensorapid and Uster Tensojet are the most common instruments used to measure single end strength and elongation. These instruments work on the CRE principle of tensile testing. Apart from single values, this instrument also calculates mean value coefficient of variation and the 95% confidence range of maximum force, tenacity, elongation, and work done.

To increase the breaking strength, core yarns are used in which the elastomeric filament such as spandex in the core is used to provide sufficient elasticity while the outer layer is made up of staple fibers of natural cotton fibers or any other. In this way, breaking strength is improved. It was also found that properties such as yarn strength and elongation of carded compact yarns are even better than conventional combed yarns.

4.6 Yarn Friction

Yarn friction is a very important property in knitting as the yarns are rubbed against various parts of machine during their passage from the creel to the needle. A yarn's knit-ability depends on its frictional properties (i.e., yarn-yarn friction and yarn-metal friction with regard to the knitting needles and generation of fluff). So, lubrication is an important process, which plays vital role in reducing the friction. During knitting yarn passes through various guides and tensioners; as a result, the frictional forces between yarn and those surfaces can cause yarn breakages due to which yarn lubrication is very much necessary to achieve the better-quality knit products. Yarn lubrication tends to ease the knitting process by offering the better loop formation. Yarn lubrication is of prime concern in case of synthetic yarns to counter the issue of static charges and helps in the dissipation of static charges. Yarn lubrications help in keeping the yarn structure intact by increasing the cohesion of fibers or filaments together. Paraffin waxes are used as lubricant for knitting yarn. Waxing is the process which is almost exclusively used in all automatic and manual winding machines for yarns which are meant for knitting. Proper selection of the suitable wax rollers should be according to the requirement of the type of yarn and fiber.

It is suggested that on average the wax concentration should be approximately 2 grams per kg of yarn.

I. Cotton and blends of cotton – 1.0 to 2.0 g/kg
II. Man-made synthetic yarn – 0.5 to 1.5 g/kg
III. Wool and blends of wool – 2.0 to 3.0 g/kg

In addition to above parameter, multiple other parameters are also important. To achieve high-quality knit products, proper yarn winding and management are very necessary.

4.7 Yarn Winding

Better yarn winding ensures the better quality of yarn and its better performance during knitting. Yarn winding has a very significant effect on the yarn tension which ultimately affect the yarn tension during knitting. Low yarn tension is another problem which leads to lower the take up of wax. The wax roller orientation and shape also affect the wax application on yarn. The wax take up can be random due to these conditions. Moreover, yarn winding has very prominent impact on the yarn twist. Also, the different winding angle of the yarn package has effect on the ultimate properties of the yarn during knitting. This can affect the knitted fabrics quality such as skewness, Bias, and distortion of the knitted fabrics.

4.8 Yarn Management/Handling

Effective measures taken for the yarn management result in better knitting efficiency, fabric quality, and reproducibility of knitted products; one of these parameters is the *package handling*. Effective handling of yarn ensures the higher quality. A poor handled package can affect the knitting process adversely even with the proper waxing. The yarn package and its protection are a very critical factor which play vital role in maintaining the yarn quality for the end processes and product. Some spinners enclosed each yarn package into polyethylene bags, or the yarns are wrapped in tissue paper. The plastic bag is far more effective when it properly encloses the yarn to avoid direct contact of yarn surface with the contamination of any type. If polyethylene bags are used to wrap the individual package, make sure they fit the yarn package tightly enough to prevent the movement of yarn within the bag, which could result in the outer layers of yarn sloughing-off the package. Do not remove the yarn package from the bags until the yarn is placed on a creel or a peg truck to be used in the knitting department.

Another parameter can be the *conditioning*. The conditioning process increases the friction of the waxed yarn which can lead to complication. Moisture increases the friction of the yarn. In addition, if too-damp packages are used in knitting, poor waxing occurred because the high moisture in yarns take little wax. Bags also tend to be beneficial in holding in the moisture from previous humidification of the yarn package.

Shipping of yarn among knitting units is done by trucks. In very rare cases, it is done via rail. In course of shipping, the temperature inside the truck containers become high which influence the yarn quality and knit-ability. This causes the loss of moisture from yarn. Another effect of mishandling of yarn during shipping is decaying and retrogression of paraffin waxes. As heating process can affect the wax on the yarn. The heat melts the wax and yarn absorbs it. The wax goes to the internal structure of yarn leaving the surface where it has no use. It has a severe effect on the frictional characteristics of yarn which ultimately affects the quality of knitted fabrics. During this, there is chance of mishandling the yarn packages which causes the damages to the yarn packages during loading and unloading.

Containers, pallets, or cartons must be properly stored in yarn inventory/warehouse. It is necessary that the yarns must be stored in a neat, secure, and dry area with easy access and retrieval. Yarns should never be mixed by supplier, lot, or shipment date. Yarns should never be left uncovered when in containers or trucks [15].

5 Yarn Quality Standards for 20/1 and 30/1 Cotton Combed Yarn

It is very important to maintain standard for the achievement of better quality. Each knitting industry should have the standard targets to check whether the incoming yarn quality is up to the mark or not. It will improve their efficiency and reduce the rejection. Quality standard / technical requirements for the cotton (carded and combed) yarns of different counts for knitting are given in Tables 5 and 6 for reference.

Table 5 Recommended technical requirement for cotton carded yarns for knitting

Count	Ne16/1	Ne20/1	Ne24/1	Ne30/1
Type/quality	Carded	Carded	Carded	Carded
End use	knitting	Knitting	Knitting	knitting
Count CV%	1.35	1.35	1.35	1.50
CLSP	2400	2350	2300	2230
TPI	14.80	16.50	18.61	20.81
TPI CV%	2.50%	2.50%	3.00%	3.00%
TM	3.70	3.70	3.80	3.80
Twist direction	Z	Z	Z	Z
Uster%	10.0	10.75	11.20	11.50
Thin/km (−50%)	0	1	2	3
Thick/km (+50%)	35	65	115	165
Neps/km (+200%)	58	85	135	215
Total Imperfection	93	151	252	383
Hairiness (−)	8.70	8.10	7.60	6.90
Average RKM	16.00	15.50	15.30	15.00
RKM CV%	8.50	8.50	8.50	9.00
Elongation%	5.10	4.80	4.60	4.30
Elongation CV%	8.50	8.50	8.75	9.00

Table 6 Recommended technical requirement for cotton combed yarns for knitting

Count	**Ne16/1**	**Ne20/1**	**Ne24/1**	**Ne30/1**	**Ne40/1**	**Ne50/1**
Type/quality	**Combed**	**Combed**	**Combed**	**Combed**	**Combed**	**Combed**
End Use	knitting	Knitting	Knitting	Knitting	Knitting	Knitting
Count CV%	1.25	1.25	1.25	1.35	1.35	1.35
CLSP	2550	2500	2450	2450	2400	2400
TPI	14.00	15.62	17.39	19.71	22.76	25.45
TPI CV%	2.50%	2.50%	2.50%	2.50%	3.00%	3.00%
TM	3.50	3.50	3.55	3.60	3.60	3.60
Twist direction	Z	Z	Z	Z	Z	Z
Uster%	7.80	8.30	8.70	9.50	10.25	10.50
Thin places (−50%)/km	0	0	0	0	1	2
Thick places (+50%)/km	4	8	13	16	30	40
Neps (+200%)/km	9	10	20	28	55	60
Total Imperfection	13	18	33	44	86	102
Hairiness (−)	8.50	7.90	7.50	6.70	5.80	5.30
Average RKM	16.50	16.50	16.30	16.00	15.80	15.70
RKM CV%	8.00	8.00	8.50	8.50	8.50	8.50
Elongation%	5.40	5.00	4.80	4.60	4.30	4.40
Elongation CV%	8.00	8.00	8.50	8.50	9.00	9.00

6 Yarn Testing

Yarn performance are identified and measured so that an appropriate yarn is used to produce fabric. There are some testing's done for the evaluation of yarn quality parameters which help in the selection of ideal yarn for end purposes. Standard test methods determine yarn parameters like yarn count, twist, strength, etc. Some of them are mentioned in Table 7. Different standards are developed by organizations like American Society for Testing and materials (ASTM) and International Organization for Standardization (ISO).

7 Knitting Yarn Defects

Yarn is one of the important factors affecting the quality of knitted fabrics. Fault in knitting occurs due to main three reasons. These are (1) faults due to knitting environment, (2) faults due to knitting process, and (3) fault due to yarns. In this section we shall discuss the description of the common defects which occurs during knitting process due to yarns. These defects causes and possible remedies. Yarn faults named as *end breakages*, *slubs*, and *knots* contribute in 70% of all defects of circular knitted fabrics. *Knots* in spun yarns for knitting are the main cause of end breakages and machine stoppage which effect the machine efficiency and knitted fabrics quality. End breakages result in hole in knitting fabrics most of which are near take

Table 7 Testing standards

Yarn	Description	Standard test method
Single yarn	Yarn count by short length	ASTMD-1059-97
Single yarn	Yarn count by skein method	ASTMD-1907-97
Single yarn	Yarn count by automatic tester	ASTMD-6587-2000
Single yarn	Yarn variability by automatic tester	ASTMD-6612-2001
Single yarn	Twist by untwist retwist method	ASTMD-1422-99/ ISO 17202
Single yarn	Twist in yarn by direct counting	ASTMD-1423-02 / ISO 2061
Single yarn	Yarn Evenness	ASTMD-1425-96
Single yarn	Yarn appearance	ASTMD-2255-02
Single yarn	Yarn classifying and faults	ASTMD-6197-99
Single yarn	Yarn hairiness by photoelectric	ASTMD-5647-01
Single yarn	Yarn to yarn abrasion resistance	ASTMD-6611-00
Single yarn	Yarn coefficient of friction to solid material	ASTMD-3108-01
Single yarn	Yarn coefficient of friction to yarn	ASTMD-3412-01
Single yarn	Determination of yarn construction	ASTMD-1244-98
Single yarn	Test method for yarn crimp	ASTM D-3883-99
Single yarn	Determination of yarn shrinkage	ASTMD-2259-02
Textured yarn	Bulk properties of textured yarn	ASTMD-4031-01
Textured yarn	Stretch properties of yarn by skein method	ASTMD-6720-01
Textured yarn	Crimp and shrinkage for textured yarn	ASTMD-6774-02
Cotton yarn	Moisture of cotton by oven dry method	ASTMD-2495-01
Single yarn	Tensile properties by single strand method	ASTMD-2256-02/ ISO 2062
Single yarn	Yarn breaking strength by skein method	ASTMD-1578-93/ ISO 6939

downside as the knots cling in the spaces between the needles. End breakages is greatly affected by machine gauge, cam angle, and coefficient of friction of yarn. Yarn breakage is directly proportional to machine gauge and coefficient of friction while inversely proportional to cam angle.

Drop Stitches (Hole)

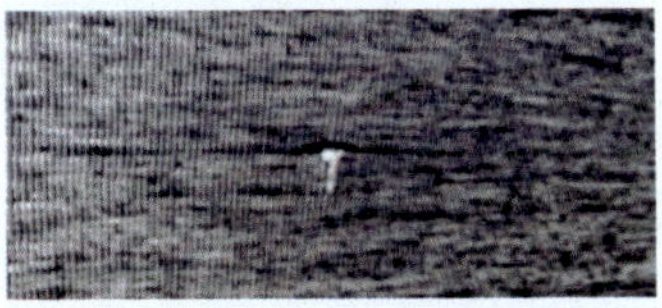

Causes:

- High yarn tension
- Yarn overfeeds or under feed
- Obstructions/obstacles in yarn passage
- Slubs, neps, knots in the yarns

Remedies:

- Uniform yarn tension and yarn feed rate
- Removal of any contaminations or fluff in yarn guides or eyelet

Barriness

Causes:

- Yarn count variation
- Mixing of the yarn lots
- Variation in hardness of yarn package

Remedies:

- Keeping uniform yarn tension on all feeders
- Must ensure that yarn used for knitting belongs to same lot
- Hardness of all yarn package should be uniform

Streakiness
Causes:

- Faulty winding of yarn packages
- Slippage of yarn from the belt and IRO pulley
- Obstruction in yarn path/passage

Remedies:

- Ensure uniform yarn path
- Removal of winding faults. Ensure flawless yarn winding
- Smooth running of yarn on belt and IRO pulley. Avoid yarn slippage

Snarls
Appearance of big loops of yarn on the surface of fabrics

Causes:

- Unbalance yarn twist or high twist in yarn

Remedies

- Using yarns with recommended TPM only

Contamination

A particle of different materials

Causes:

- Inappropriate cleaning of raw materials
- Packing bags contamination
- Different materials knitting on the same place

Remedies:

- *Care should be taken during cleaning and knitting*

Spirality

Spirality is the appearance of distorted wales of the fabrics. Usually, this distortion is in diagonal shape

Causes:

- Due to high TPM of yarn

Remedies:

- Use of yarn with recommended TPM

Pilling

Pilling is the accumulation of small balls of fibers on the surface of fabrics.

Causes

- Low yarn quality
- Yarns with more hairiness and higher number of protruding fibers

Remedies

- Use of yarn of higher quality
- Use of yarns with less hairiness

References

1. E. Wood, 19. Principles of yarn requirements for knitting key terms and concepts. Wool Process, 1–12 (2009)
2. H.M. Behery, Yarn structural requirements for knitted and woven fabrics. Adv. Yarn Spin. Technol., 155–189 (2010). https://doi.org/10.1533/9780857090218.1.155
3. H. Jamshaid, Spirality in knitted fabric. J. Text. Sci. Eng. **8**, 350 (2018). https://doi.org/10.4172/2165-8064.1000350
4. D.H. Black, Knitting with cotton and cotton blend open-end spun yarns. Text. Res. J. (1975). https://doi.org/10.1177/004051757504500109
5. H. Hasani, An investigation into the effect of fabric structure and yarn twist direction on the curling behavior of single jersey weft knitted fabrics. J. Fash. Technol. Text. Eng. **2**(1), 1–6 (2014). https://doi.org/10.4172/2329-9568.1000107
6. A. Primentas, C. Iype, Spirality of weft knitted fabrics: Part III – An innovative method for the reduction of the effect. Indian J. Fibre Text. Res. **28**(2), 202–208 (2003)
7. M.D.d. Araujo, G.W. Smith, Spirality of knitted fabrics: Part I: The nature of spirality. Text. Res. J. **59**(5), 247–256 (1989). https://doi.org/10.1177/004051758905900501
8. N. Özdil, A. Marmaralı, S.D. Kretzschmar, Effect of yarn properties on thermal comfort of knitted fabrics. Int. J. Thermal Sci. **46**, 1318–1322 (2007). https://doi.org/10.1016/j.ijthermalsci.2006.12.002
9. S.H. Sanad, H.M. mahmoud, M.A. El-sayed, Production of carded compact cotton yarn of comparable quality to the combed conventional ring yarn. Egypt. J. Agric. Res. **89**(1), 203–212 (2011). https://doi.org/10.21608/ejar.2011.173984
10. D. Yilmaz, M.R. Usal, A comparison of compact-jet, compact, and conventional ring-spun yarns. Text. Res. J. **81**(5), 459–470 (2011). https://doi.org/10.1177/0040517510385174
11. Y. Elmogahzy, *Structure and Mechanics of Yarns. Structure and Mechanics of Textile Fibre Assemblies*, vol 1, 2nd edn. (Elsevier Ltd, 2019). https://doi.org/10.1016/B978-0-08-102619-9.00001-8
12. J.R. Wagner, E.M. Mount, H.F. Giles, Monofilaments. Extrusion, 585–591 (2014). https://doi.org/10.1016/b978-1-4377-3481-2.00051-x
13. C. Lawrence, Fibre to yarn: Filament yarn spinning, in *Textiles and Fashion: Materials, Design and Technology*, (Elsevier Ltd, 2014). https://doi.org/10.1016/B978-1-84569-931-4.00010-6
14. Solutions, Texcoms Textile, Textile spinning. Texcoms Text. Solutions **II**, 7–28 (2019)
15. Technical Bulletin. "Cotton spun yarns for knit and woven fabrics" © 2003 (Cotton Incorporated)
16. X. Wang, Fundamentals of yarn technology. Deakin University, Geelong, Victoria 3217, Australia, (2000).
17. C. Rameshkumar, P. Anandkumar, P. Senthilnathan, R. Jeevitha, N. Anbumani, Comparitive studies on ring rotor and vortex yarn knitted fabrics. Autex Res. J. **8**(December), 100–105 (2008)
18. P.R. Lord, P.L. Grady, The twist structure of open-end yarns. Text. Res. J. **46**(2), 123–129 (1976).
19. Y. Kim, Y. Ryul, W. Oxenham, Analyzing physical spun. Text. Res. J. **72**(2), 156–163 (2002)

Weft Knitting Machines

Awais Ahmed Khan, Rajesh Mishra, and Hafsa Jamshaid

1 Introduction

Knitting is the process of fabric formation by producing series of interlinked loops. It can be used to create a garment or some other type of clothing. Knitting industry has grown into masses. Every human being, from toddlers to kids, boys, girls, men, and women, is using knitted products. Knitting is a vast field and comprises of two main techniques i.e., warp knitting and weft knitting. As we have discussed in previous chapters, weft and warp knitting fabric manufacturing techniques are different enough from each other in the respect of fabric properties, material used, and also by the aspect of machines. As knitting technique is divided into two types of warp and weft knitting, in the same manner machinery is also divided into two categories: weft knitting machines and warp knitting machines. The scope of this book is only weft knitting, so emphasis shall only be on weft knitting. Weft knitting has two major classifications, i.e., circular and flat knitting machines. Weft knitting can be done by hand or machines. In hand knitting, a yarn is looped onto one needle and the other needle is inserted into the stich, and by managing the positioning of needles, the new stich is taken off onto the second needle. This process is repeated to develop a garment. While in case of machine knitting, many needles are used to develop the fabric. Weft knitted fabric is developed by one set of continues yarn that travels around the fabric in circular knitting, while across the fabric in flat knitting techniques respectively.

A. A. Khan (✉)
Faculty of Textile Engineering, National Textile University, Faisalabad, Pakistan

R. Mishra
Faculty of Engineering, Czech University of Life Sciences Prague, Prague, Czech Republic

H. Jamshaid
School of Engineering and Technology, National Textile University, Faisalabad, Pakistan

H. Jamshaid, R. Mishra (eds.), *Knitting Science, Technology, Process and Materials*, Textile Science and Clothing Technology,
https://doi.org/10.1007/978-3-031-44927-7_3

2 Circular Knitting Machines

Circular knitting machines have circular-shaped frame on which the whole assembly is mounted. As the name suggests, circular knitting machines include all those machines in which needles bed are in circular shape irrespective of the type of needles used. These machines produce circular-/tube-shaped fabric in different sizes/diameters depending on the field of application. Needles are arranged in circular order along the perimeter of cylinder or dial. These machines are used to make yardage, sweater bodies, socks, etc. These machines have been designed and manufactured for mass production of knitted fabrics. The history of circular knitting machines dates 200 years back in which a number of researchers contributed to the development of machine technology. A number of names are included in the list of inventors but among all of the inventors it is believed that patent of Decroix is the first considered for the invention of circular frame in 1798. Later on in 1849, *Mosses Mellor* developed a circular revolving frame in which bearded needles were arranged vertically. It was found that in the same year 1849, use of latch needle was patented by *Matthew Townsend* [1]. Baiyuan Machine (China), Mayer & Cie (Germany), Terrot (Germany), Santoni (Italy), and Fukuhara (Japan) are the major companies operating in Global Circular Knitting Machine Market. The Global circular machinery market is expected to grow at a CAGR of 4.0% over the forecast period of 2022–2027 [2].

https://www.mordorintelligence.com/industry-reports/global-circular-knitting-machine-market#].

3 Classification of Circular Knitting Machine

In this chapter for the proper understanding of the reader, circular knitting machines are divided into sub-categories. The classification of circular knitting machines is done based on the number of needle beds and on diameters which is illustrated in Figs. 1 and 2.

- Classification based on needles bed
- Classification based on diameter

3.1 Classification Based on Needles Bed

Weft knitting machines are categorized w.r.t. many aspects but a major classification aspect of weft knitting machines is w.r.t. number of needle beds and their arrangement. On the basis of number of needle bed, weft knitting machines are divided into two types:

- Single knit/needle bed
- Double knit/needle bed

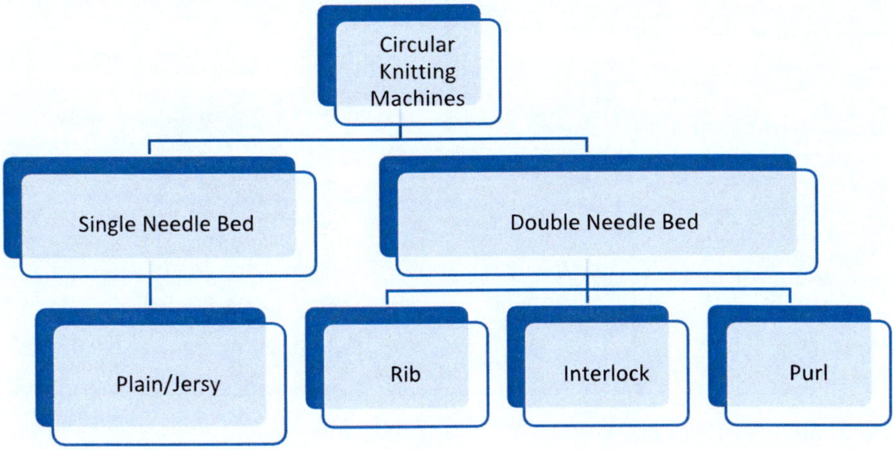

Fig. 1 Classification of knitting w.r.t. needle bed

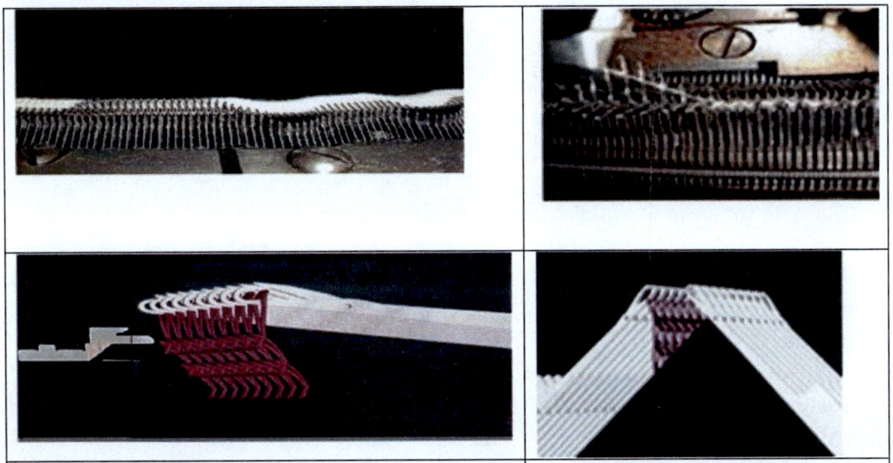

(a) Single Jersey (b) Double Jersey

Fig. 2 Description of machines

4 Single Needle Bed

Single needle bed machines are usually known as single jersey machines. In single jersey machines, there is only one cylinder along which the needles are arranged (as shown in Fig. 3) to produce tubular fabric. As there is only one set of needles in single jersey machines, single jersey fabrics comprises only one layer of the knitted fabric. Needles are arranged on the vertical cylinder in grooves present around the circumference of cylinder. Single jersey machines utilize the set of sinkers for the

Fig. 3 Single jersey machine

production of fabric. Sinkers hold the loops and perform major role in loop formation and casting of loop. Single jersey machines limit the use of woolen yarn upto 20 gauge or coarser because these gauges can easily use double-folded woolen yarns.

In single jersey machines, the fabric formation is supported by the latch needles and sinker's ring upon which sinkers are placed. In single jersey machines, usually the cam system is stationary while needles and sinkers revolve around it. Sinkers play an important role in loop formation. Mainly two types of sinkers are used which are explained in the next section. In single jersey, the operation of loop formation with the help of sinkers is commonly known by the term *Contra knitting* or *Relative technology*. This describes the relative motion of needles and sinkers.

Baiyuan (Fujian **Baiyuan** Machinery Co, Ltd, China), leading manufacturer of knitting machines, developed high-speed single jersey machines with a wide range of diameter 11 inch to 68 inch and gauge of 9-52. 33-218 feeders can be utilized in this machine. Another model having the exchangeable knitting assembly which facilitates to switch to fleece or terry. This model also has the reverse platting technology with 30–34-inch diameter and 14–24 gauge. The feeder capacity of this machine is 60Feeder/68Feeder [3]. Terry or fleece knit machines also come in the category of single jersey machines. In these machines, two yarns are used to knit fabrics with the help of specialized sinkers. Sliver knitting machines are also single jersey knitting machines.

5 Double Needle Bed

Double needle bed is widely known as double jersey machines. These machines have double needle bed, a cylinder, and a dial. There is an extra needle set which is placed horizontally on the dial above the vertical cylinder. Dial needles are adjacent to the cylinder needles. There are two types of placements and operation of dial needles relative to the cylinder needles. There is much correspondence between the dial and cylinder needles position and operation which result in different types of fabrics. Based on the relative needles position, double jersey machines are further divided into two categories: Rib and Interlock, which are briefly described below. The diameter of dial and cylinder is same. *Needle timing* is a very important term that defines the difference between Rib and Interlock. Needle timing describes the relationship between position of dial and cylinder needles and their course of action. In rib and interlock machines, the knitting action is performed according to the delayed timing. Prior to the describing the difference between the interlock and rib knitting machines, there is a need for a proper understanding of the term known as needle gating. Needle gating refers to the relative arrangement of needle of the dial and cylinder. In interlock gaiting, the needles or dial are placed exactly opposite to the needles of a cylinder. Whereas, on the other hand, in the rib gating unlike the interlock, the needles are placed in an alternative order. This arrangement of the needles permits the operation of both needles at the same time. There are three basic types of double bed machines as described in the below section.

Rib

It is found in the literature that *Jedediah Strutt* was the first to invent the Rib frame in 1755. He was first who utilized the second set of needles. In rib knitting, needles of cylinder and dial are arranged at right angle to each other. During fabric formation, the loops are drawn in opposite direction so that face and back loops alternate in each course. Needle gating is the term used for relative arrangement of dial and cylinder needles. Needle gating of rib knitting machine is shown in Fig. 4.

A leading manufacturer of knitting machine SINTELLI (Xinda Precision knitting machine) presented its latest rib circular knitting machine model DJR. Which has the capacity of conversion of machine setting from rib to interlock. Machine is commonly used for the production of leisurewear. It offers a wide range of raw materials to be utilized for the production of knitted products including cotton, silk, synthetic fiber, artificial wool elastic cloth, or mesh [4].

Interlock

Interlock machine is a complex machine offering diversity in design and a balanced fabric structure. Interlock machine is also the double bed machine having special arrangement of needles to produce interlock fabrics. In interlock machines, the

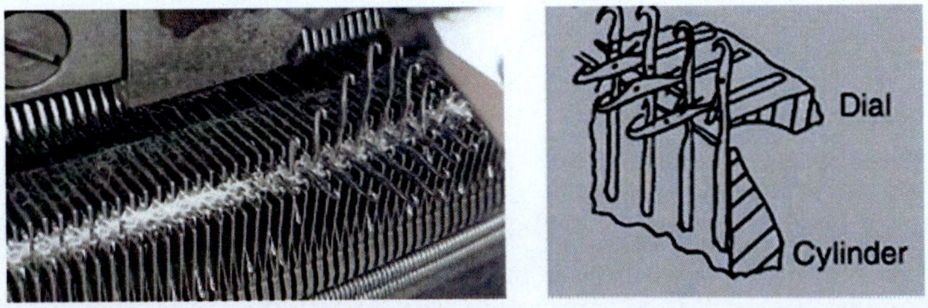

Fig. 4 Rib gating of machine needles b Knitting technology a comprehensive handbook and practical guide Third edition David J spence

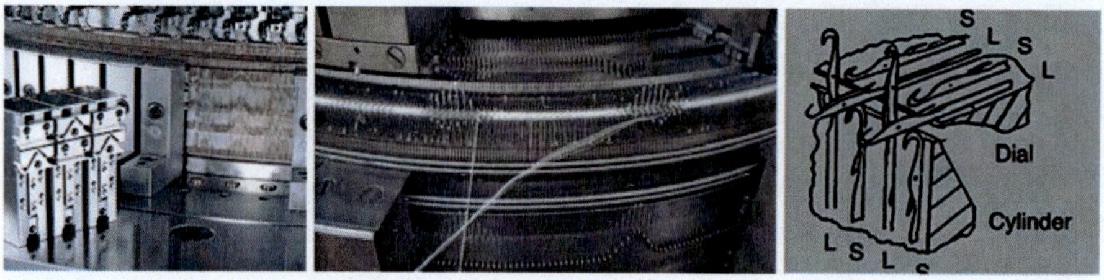

Fig. 5 Interlock gaiting of machine needles c Knitting technology a comprehensive handbook and practical guide Third edition David J spencer

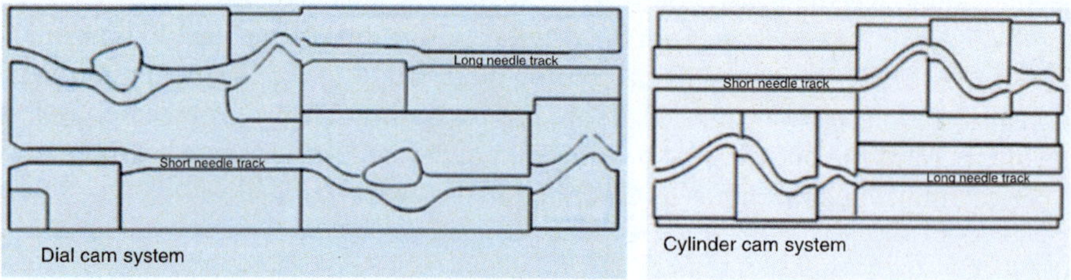

Fig. 6 (**a**) Dial cam system of knitting machine. (**b**) Cylinder cam system of knitting machine. Knitting technology a comprehensive handbook and practical guide Third edition David J spencer (Reproduced)

needles of dial and cylinder are arranged exactly opposite to each other. In interlock knitting machine, rather than arrangement of same set of needles exactly opposite to each other needles are placed in alternative arrangement. The loops produced by one set of needles are interlocked by the other set of needles. In interlock knitting machines, as there are two sets of needles, i.e., short and long needles, at least two yarns are required for the fabric formation, so separate yarn is supplied to each set of needles. The loops produced by the two sets of needles (long and short) are produced with a time gap as shown in Fig. 5. The same set of the needles of both beds form loops with conjunction with each other, like the long needles of one bed form loops in collaboration with the short needles of other bed. That's why only alternate needles knit at a feeder and that's why interlock machines have the capability to operate in the finer gauges. The productivity of interlock machines is half, due to limited accommodation of number of feeders. Figure 6 shows the dial and cylinder cam system of double-knit machines.

6 Purl Knitting Machines

This is also double bed knitting machine. In purl knitting machine, special types of needles are used for the formation of loops. Lath needles with two hooks are used which facilitate the transfer of loops from one bed to the other. This leads to the utilization and requirement of only one set of needles. Purl knitting machines offers diverse range of derivative of rib and interlock by changing the needle gaiting of the machine. Two different structures can be produced on the either side of the fabrics. Purl knitting machines are very rare.

6.1 Classification Based on the Diameter

Diameter is a very important parameter which has a significant effect on the final knitted products. Circular knitting machines produce tubular fabrics in variable diameter for a variety of applications. In case of circular machines, operating width is measured in from the first groove of the needle bed to the last groove. Width of circular machines is the diameter of cylinder which is measured in inches. The diameter is measured between the two opposite needles.

Machines are classified into three different categories based on the diameter of the needle bed, which are briefly described in the next sections and illustrated in Fig. 7.

- Large diameter machines
- Medium diameter machines
- Small diameter machines

Socks knitting machines are also circular knitting machines, but they fall in the category of small diameter range. While body size undergarment and garments are

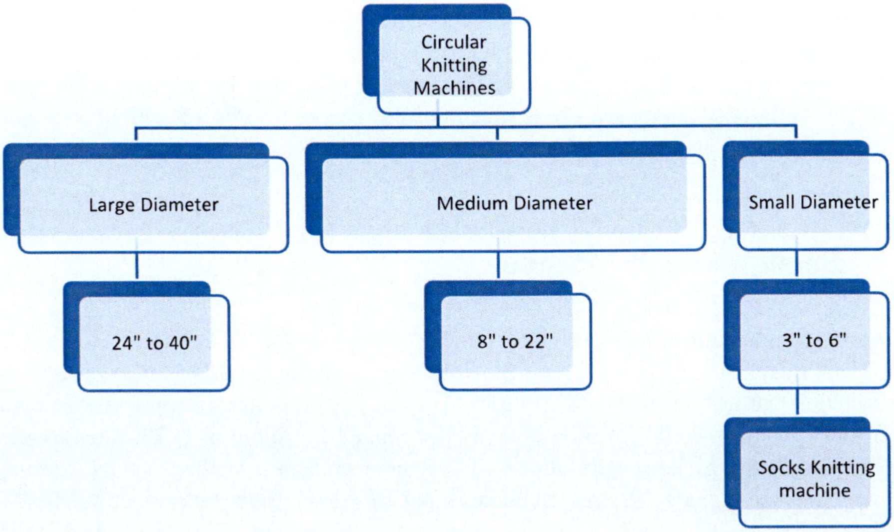

Fig. 7 Classification of knitting machines w.r.t. diameter

mostly manufactured on the medium diameter knitting machines. Weft fabric for making apparel by cut and sew process are made by the large diameter range machine.

7 Large Diameter Machines

The diameter of machine decides the number of systems involved in knitting fabric. So, by increasing the diameter of machine, the systems involved in the development of fabric increase. The increase in diameter results in increase in the number of courses produced per revolution of the cylinder [5].Circular large diameter machines are used to produced endless tubular fabric for applications in mass production like outwear (T-shirts, sleepwear, sportswear).

Santoni, Italy based, the manufacturer company of knitting machines developed latest single jersey machine with very fine gauges. There are two versions of ATLAS single jersey circular machines. The Single jersey Atlas have machines with fine gauges up to 50 and 60 needles per inch. The latest version of this machine comprises of special types of under jacks with development in cylinder. The machine has special sinker-less technology for the knitting operation. The circular knitting machine with ultra-fine gauge of 80 and large diameter 30 inch was presented by Santoni at ITMA Barcelona having a total number of needles 7536 in cylinder. The unfinished fabric width of 220 cm was able to be produced.

8 Medium Diameter Machines

Circular knitting machines having diameter of 8 to 22 inches are called medium diameter circular knitting machines. These machines can be available in single jersey as well as double jersey. These machines can produce body sized tubular fabrics along with the capability to produce the welts with separate yarn.

9 Small Diameter Machines

Small diameter machines are also called *Hosiery machines*. The term Hosiery is mostly misused and applied on all the knitwears; contrary to this, Hosiery is a term used only for the garments utilized for the covering of legs and feet. The first known hosiery machine with power source was produced by *Shaw* in 1879; later on the year 1887, the pickers were added to knit heel and toe pouches. Socks knitting machines are included in small diameter machines (Fig. 8). In socks machine, firstly the tubular part is produced, then the foot and toe at the end. Toe closing is done at the end.

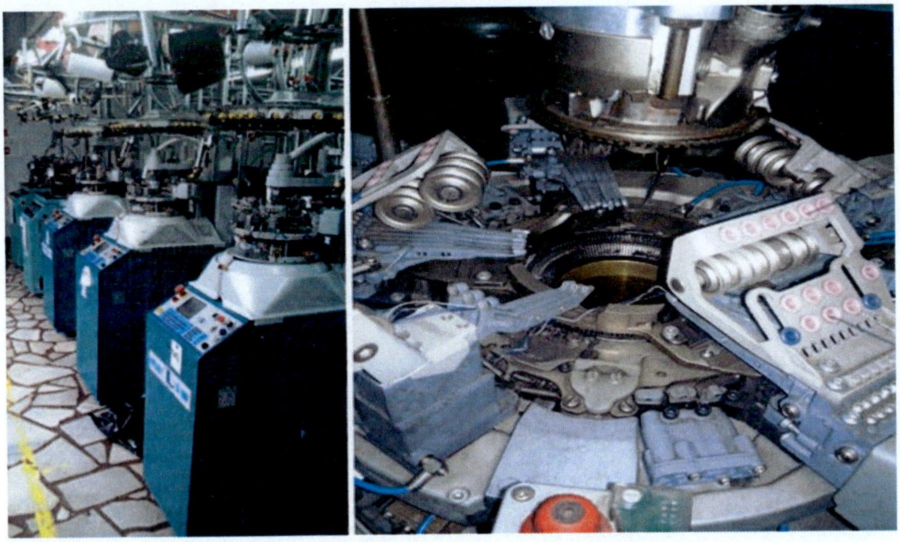

Fig. 8 (**a**) Socks knitting machines. (**b**) Close-up view of the knitting head of socks machine

In small diameter machines or Hosiery machines, the gauge is expressed as diameter and total number of needles, for example, 4-inch × 400 needle. Small diameter machines basically produce tubular fabric with small diameter. Contemporary developments in the field of small diameter hosiery machines, specifically single cylinder and fine gauge, are centered in Italy. One of the most well-known companies for socks manufacturing machines is Lonati *(Italy).* These machines have electrically controlled operations which enable the variation in stitch length during different processes such as toe closing and finishing operation of the product. Mostly the small diameter hosiery machines comprise of a revolving cylinder with the only exception of *Griswold* type machines. Small diameter knitting machines can produce seamless garments including shirts, swimwear, underwear, and many more.

The Hosiery machines are classified into three types based on the cylinder arrangement

Single/Mono Cylinder Machines

Single cylinder machines usually produce ladies' seamless hose and tights that are knitted in basic plain structure with fine gauges. Holding down sinkers are utilized in these types of machines.

Double Cylinder Machines

Double cylinder machines offer the broad rib or purl knit structures. These structures are used in men's, children's and lady's half hose and socks.

Cylinder & Dial Machines

These machines are also called *True-rib* Machines. As the rib socks are knitted on this type of machine. In these machines, half number of needles are placed in dial and the half number of needles are placed in cylinder. The needles of both dial and cylinder are in between each other like each second needle on the dial is opposite to the cylinder needle. These machines have the limitation which restrict the formation of broad ribs like 6 × 3 Rib. The gauge of hosiery machines is expressed as diameter and total number of needles; for example, the machine having 4-inch diameter and 400 needles in single cylinder will be expressed as 4-inch × 400 needles.

10 Major Parts of Circular Knitting Machines

10.1 Frame

The shape of the frame of knitting machine is in accordance with the needle bed either circular or flat. Machine frames provide support to the machine elements and mechanisms. In a circular knitting machine, frame is the platform at which all the members/elements of knitting machine including the active members and passive members of the machine are placed in successive order. The part of frame which is the lowermost comprise of the legs and the other lower crossmembers. Above this is the upper subframe which holds or accommodate the legs of immediately upper subframe rest. The immediate upper subframe contains the legs and upper members hold a platform with its relative ring. The rest of the subframes have similar architecture. It is noted that the sides of the lower most part of frame or subframe and resting above it the immediate upper subframe form a free opening which provide direct and easy access to the internal zone of the machine. The frame provides plate form for the arrangement of different assemblies of knitting machine. Various members of machine are distributed among the different stories of frame or on the subframes which are placed on one another. The subframes have legs. The top and intermediate subframes are arranged to support a flat horizontal platform which have circular opening built on the edge of the ring which provide support for the members of machine [6]. The machine frame must have enough strength to absorb the acceleration and various forces to which knitting machine is subjected during different operations. The frame of circular knitting machines comprises of different parts. Lower frame which has cross beams and feet which provide support for fabric takedown. Secondly upper frame with carrier plate, cylinder needle bed along with cylinder bearing columns, and dial needle bed housing with dial bearing.

10.2 Drive

Drive is the primary element for the performance, operation, or working of the any machine. Drive of the knitting machine has absolute impact on the efficiency of the knitting machine and the quality of the fabric thus formed. Drive of the knitting machine seek utmost importance with the increase in the functional attributes of the knitting machines. Increasing the productivity and design versatility demands greater drive reliability. There are mainly two power sources for the knitting drive which is a manpower and electrical power. Initially the hand knitting machines driven by manpower. With the advancement in the field of technology electrical motors were introduced in the knitting machines also. Motor-driven or electrical power-driven machines result in higher efficiency. Depending on the machine diameter and feed system, the motor of compatible horsepower is installed in the knitting machines. Brake and clutch arrangements are used to connect motor pulley with machine pulley. The gearing arrangement is used for the transfer of motion from machine pulley shaft to the cylinder. If there are two cylinders or dial and cylinder assembly with the help of gearing system, motion is transferred from one cylinder to another. There is a synchronization in the drive mechanism and rotational motions of the machine. This synchronization is achieved with the help of gearing system which is used to deliver power to the PIV pulley of yarn feed system from the rotating cylinder. Spreader/Stretcher board also gets it drive from the rotational motion of cylinder. Moreover, eccentric mechanism is used to transfer the motion of stretcher board to the takedown rollers and cloth rollers. In some cases, for the takedown rollers, an additional power system is used.

10.3 Yarn Feeding System

Yarn feeding system includes different elements or parts which ensure the proper feeding of yarn to the needles. Yarn feeding system mainly comprises yarn package, tensioning devices, yarn feed control, and yarn guides (Fig. 9). Based on yarn feeding system, basically there are two system and mechanism of yarn feeding machines are classified into two categories.

Moving Cam System with Stationary Needle Bed This type of feeding system is found in the small diameter knitting machines in which there is single feed. As the needle bed is stationary in this type of machines, the fabric is also stationary. Flatbed

Fig. 9 Positive yarn feeding system. Thread guide

knitting machines come in this category. The moving cam system known as *Carriage* is in traverse motion along the length of stationary needle bed.

Stationary Cam with Revolving Needle Bed In this type of machines, usually there is multiple feed system. Owing to the moving needle bed, fabric thus produced also rotates. The yarn packages in this system also must rotate to avoid yarn entanglement. Based on principle of yarn feeding, there are two types of yarn feeding system which are explained below.

Negative Yarn Feeding
In negative yarn feeding system, the yarn is directly drawn from the yarn packages, i.e., cones. This yarn supply is continuous. Negative yarn feeding doesn't require any extra assembly because the yarn is automatically drawn in this process. In this yarn feeding system, it is difficult to achieve uniform tension on yarn.

Positive Yarn Feeding
In positive yarn feeding, an external element is involved in yarn drawing from the package and supplied to the needles. Positive yarn feeding system of knitting machine comprises pulleys which get their drive from gears and belts as shown in Fig. 9. Positive yarn feeding supplies yarn to the needles at a constant rate and uniform tension. To achieve the positive yarn feed in knitting machine, a tape device is placed on the feeder which supply the yarn. The rate of yarn feed is variable according to the machine running speed. The rate of yarn feeding varies depending on the speed of tape device. The speed of this feeding tape device can be adjusted from Variable Dia. for Quality (VDQ) pulley.

A yarn feeding device used for the plating technique patented by *Onishi Yasushi et al,* yarn feeding device consist of one feeding device for ground yarn and second feeding device that is used for feeding elastic yarn. A guide roller which guides the elastic yarn through the path. A disentanglement device which prevent the entanglement of elastic yarn with the ground yarn [7].

11 Yarn Feeders

Yarn feeders are arranged at regular spacing around the circumference of cylinder. More number of feeders means more production. Although production can be increased by increasing machine speed, but it will increase machine vibrations, jerks, etc. Feeders are arranged at regular intervals for supplying a greater number of yarns to the needles simultaneously for achieving higher production. Each feeder produces separate course in each revolution of the machine. Maximum number of feeders on the machine depends on diameter of the machine, machine gauge, type of the machine w.r.t. single jersey or double jersey, and patterning facilities.

12 Cams

Cam plays a key role in the loop formation. During the loop formation, the movement of needle is done by the cam. Each type of needle is supported by the type of cam; for example, for the formation of knit /stitch, knit cams are utilized and the formation of tuck is endured by the tuck cam. For the miss stich, miss cam is used where needle don't simply pass through without moving up and receiving the yarn. Cams are placed in the cam box and needles move through the cams. As mentioned earlier, in some knitting machines the cams are stationary, while some machines have moving cam system/box. The knitting action of needle is basically governed by the specific type of cam. Basically, there are three well-known types of stitches used in weft knitting for the development of various designs. So mainly three types of cams used in knitting machines which are as Knit, Tuck and Miss Cams which are shown in figures in Fig. 10.

The cams convert the electrical driven rotary motion into the suitable motion of the needles. Cams also play a very important role in the designing of the knit fabrics. Commonly in single jersey knit machines, single cam track is utilized, while for the diversity in the designs multitrack is the suitable option (Fig. 11). Multitrack system offering design versatility is explained in the next section.

Fig. 10 (**a**) Knit cam; (**b**) Tuck cam; (**c**) Miss cam

Fig. 11 Multitrack for CAMS

13 Fabric Takedown

The mechanism of fabric takedown includes winding devices, fabric tensioning, and accommodation devices. Fabric takedown means the downward withdrawal of the produced tubular fabric on the machine with the help of mechanism. Fabric in tabular form goes downward for rolling into roller as flattened double layer fabric. During this process, tension variation can take place across the width of the fabric which leads to wrinkle, creese marks, and stich deformation. To overcome this problem, spreader /stretcher board are provided in knitting machine as shown in (Fig. 12a) for applying uniform tension on the fabric.

There is property synchronization between the machine cylinder and fabric take up roller (Fig. 12b) to avoid the curling of the fabrics. With the production of fabric with the growing length of the produced fabric, the tension in the knitting machines varies which also affects the needles and produce stresses on the needles. This results in needle breakages and tearing apart of fabric occurs. In the previous versions of the machines, there was problem of adjusting the take up tension while running machine and as the fabrics grows. So later on, in latest technologies with the help of the scientists and researchers (European patent) the issue was addressed. The problem was encountered and solved successfully in the latest machines, so by varying the speed of take up roller as the fabric produces it helps in balancing the tension [8]. Takedown is the withdrawal of produced fabric on to the roller commonly in downward direction. The mechanism of fabric takedown is governed according to three principles.

Dead Weight Principles

This principle of takedown mechanism is discretionary rather than continuous as in this case some dead weights are clamped with the fabric to apply downward tension to the loops. This help in casting of old loops. But in this method after some time the weight touches the ground. It is not so practical for large-scale production; it is commonly used for small diameter circular machines or flatbed knitting machines for small-scale production. The barrier in the practicing this method on a wide scope is that it doesn't apply uniform tension.

Eccentric Principle

In eccentric principle of takedown mechanism, the withdrawal of fabric from the knitting zone of the machine to the downward is done with the help of pair of fluted

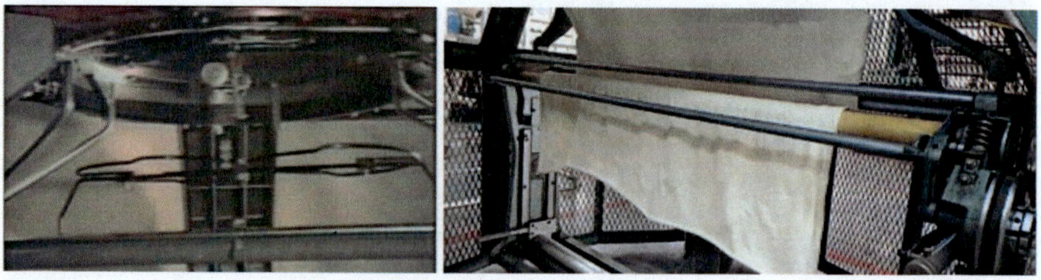

Fig. 12 (**a**) Spreader (**b**) Fabric takedown roller

takedown rollers. This principle supports the downward tension at constant rate. The name eccentric is governed to the principle which is based on the eccentric arrangement providing motion to the pair of rollers withdrawing the cloth.

Electrical Principle

In takedown mechanism, enduring the electrical principle the takedown rollers and cloth rollers are separately rotated with the help of separate electric motor. Motor is fixed on the platform which hangs below the cylinder and undergo rotational motion according to the cylinder. The tension and surface speed of the rollers can be adjusted from the electrical Control panel. It is the most widely acknowledged principle of takedown mechanism

14 Monitoring System and Service Devices

The proper monitoring of knitting process on knitting machine is very important from the quality management and efficient production point of view. The monitoring or surveillance is very important for scheduling and planning of order. The proper monitoring system helps in detection and controlling of faults. Numerous devices are being utilized for the monitoring of the faults, for example closed latch detectors, pattern recognition use for the detection of damaged needles. Moreover, capacitive sensors and optic sensors are also used for the surveillance purposes. Yarn brake tensioner is also a monitoring device which is based on the principle of gravity [9]. Knitting machine electronic panel for various controls are shown in Fig. 13.

Fig. 13 Machine control panel

During knitting process, elastane faults are very difficult to detect which affects the production. Monitoring system was developed using artificial intelligence and computer vision for the detection of fault including yarn thickness, needle defects holes, and many other. The system was developed by the San Francisco based company known as *Smartex* which claims to reduce the faults up to 0%. It was also stated that *Smartex-vi* performs real-time monitoring of production using the internet facility. It was also stated that this device is very easily installed and synchronized with the knitting machines effectively. This monitoring system tends to stop the machine on fault detection to avoid further faulty production [10]. The monitoring of knitting process has an approach of detecting faults which is based on the yarn friction and yarn input tension. It is found in the literature that the monitoring system based on yarn tension is somehow not have been in the focus. Contrarily, it plays a very important role in the production and efficiency of the knitting machine. Catarino et al. elaborated yarn tension monitoring system based on techniques involving frequency and time analysis. The yarn input tension has prominent relation with the process inside the knitting zone. So, any type of disturbance arises due to any fault of yarn, which will be represented in the monitoring system in form of wave. It is also stated that in case the cams are non-linear, the waveform represents as sinusoidal waveform. The system comprises of encoders which generate the signals for the calculation of yarn consumption and rotational speed. Strain gauge force sensors measure the yarn tension which is primary function for the detection of fault and other irregularities related to the yarn and knitting elements. Optical sensors are placed in the knitting zone and responsible for marking the spot where the revolution of cylinder started. Conditioning board interconnects all the sensors for the acquisition of data. A PC provides the platform for running different applications. The Application named *KnitLab* gathers information with the help of conditioning board and stores it in a file. The advantage of the sensors is that they do not require any additional modification for their accommodation on the machine [11]. In the literature, another application named as *Monitor Knit* detects the fault and locates the faults during the production on knitting machine. Low cost sensor for knitting machine monitoring was developed by Catarino et al. This sensor is also based on the yarn input tension. The proposed technology is based on the relationship between deflection of thin plate composed of metal and force. It was also confirmed that technology results in the better production and fault detection. It was also suggested that further developments can be made in the monitoring system of knitting machine [12].

15 Knitting Elements of Circular Knitting Machines

Knitting machines produce the loops of yarn and perform interloping of yarn to produce knitted fabric. This operation of loop formation and interloping of consecutive loops is done by different knitting elements. The knitting elements are the vital parts of a knitting machine. Knitting elements are further described in this section.

16 Needles

Needle is a very vital element of knitting machine. Needle is the first and foremost element of the knitting machine. The needles are positioned in tricks or cut of needle bed (flat or circular) at ordered intervals so that they can move easily throughout the loop formation cycle.

In single jersey machines, needles are vertically arranged on the vertical cylinder perpendicular to the ground. Generally, machine manufacturers favor to use the latch needle for their machines. The latch needle is self-actuating and no supporting part is required.

17 Latch Needles

Latch needles are mostly used in the production of weft-knitted fabrics. Latch needle was first of all patented by the Pierre Jeandeau in 1806. Latch needle is also known as tumbler needle. Knitting with latch needle was milestone in the history of knitting as before this spring bearded needle was mostly used. Latch needles eliminate the need of pressor which is commonly used for the hook closing of beard needle. Initially latch needle was adopted by the Americans while British were reluctant in adopting the latch needle. According to the British, spring beard needles produce higher quality fabrics. Later on, latch needles also proved to be better alternative. Latch needle is the most successful among the tree types of needles, i.e., spring beard, compound needle, and latch needle. Latch needles mainly consist of nine parts/attributes.

- The hook which tends to draw and facilitate in retaining the loop. Needle hook depends on the thickness of needle in other words on the gauge of knitting machine.
- The *saw cut* or commonly known as Saw Cut receives the latch of needle.
- The *cheeks* or also known as slot walls which are rivet.
- The *rivet* is punched or plain; it is used to retain the latch blade of needle.
- The *latch-blade* is primary part of latch which is situated in the latch.
- The *hook spoon* is the extension of latch blade which facilitates by bridging the gap between hook and stem of needle.
- The *stem* of the needle helps in carrying the loop during the knitting action of needle while clearing position or rest position.
- The *butt* which facilitates the needle movement in the cam. This tends to reciprocate the needle when encounter the cam.
- The *tail* of the needle is the area of needle below the butt which provide an extra support to the needle while the knitting action.

For the greater productivity, high-speed process is required. High-speed knitting process produces high stresses during the needle action, so the needles should be able to withstand the stresses and work efficiently. During the use of fancy yarns, an extra stress is applied on the hook and hook base. There should be optimum flexibility to withstand strain, and efficiently produced fancy fabrics ultimately contribute to the diversity of knitted fabrics. It is found that Groz-Beckert (Germany) the manufacturer produced special conical-shaped hook which facilitates the better thread clearance between the loop-forming elements.

Knitting with Latch Needles

Latch needles don't require extra attachment for hook closing. Yarn tension tends to move the needle latch up and down.

Knitting Operation of Latch Needles

Knitting operation of latch needles executed as the latch needles pass through different positions and loop formation is done. The knitting cycle and different positions/stages are described as shown in Fig. 14. In the start of knitting cycle, needle is in the rest/running position little bit higher than the sinker top in case of single jersey knitting. At this time, the loop which was formed on the former/last/previous feeder is enclosed in the needle hook. The second stage of knitting cycle is the opening of latch with the passage of needle through the cam path. Needle moves up with the help of clearing cam. The old loop held by the sinker slides through the needle hook encountered the latch and tend to open it. Next stage when the needle reaches at the top of the clearing cam the old loop is now cleared from the latch and at that time latch hangs down. When the needle starts to lower down following the path of stitch cam, this step lowers down the old loop under the latch at the same position new yarn is fed to the needle. There are three basic activities performed by needles: *Casting off*, *loop formation*, and *attaining idle position.*

In the figure, latch needle has been shown at different stages of the knitting cycle or loop formation cycle and also cam track or cam profile is illustrated in accordance with all the positions of knitting cycle are mentioned below in detail.

1. *Run-in*

As from the figure, it is clear that in first step needle is getting into the cam tracks area which works mainly for loop formation. Basically, this is the starting point. Needle will rise slowly and goes higher in its position.

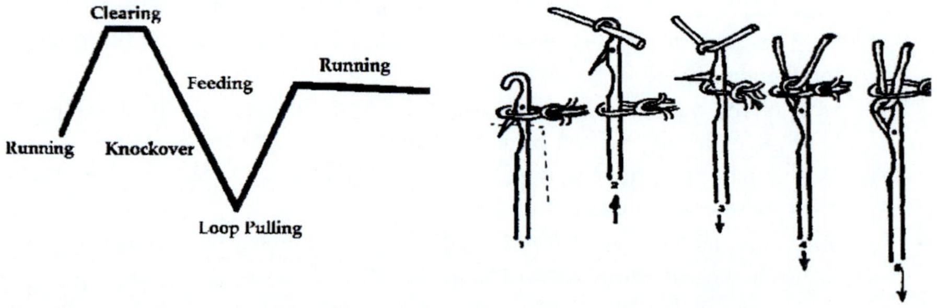

Fig. 14 Needle motion in cam and needle movement (Fundamentals and Advances in Knitting Technology, Sadhan Chandra – Woodhead Publishing 2011)

2. *Latch Opening and Clearing*
When needle butt moves upward in the cam track, then yarn loop present in needle hook moves downward and contacts the latch due to which latch starts to opening. As needle attains more height and reaches the maximum height of cam profile, then yarn loops completely come down on the stem of the needle and hook becomes free of loop and latch opens.

3. *Yarn Feeding and Latch Closing*
After clearing the old loop, needle starts to descend and hook is fed with new yarn for new loop. As needle descends, the older loop comes under the latch, and when needle moves more downward, the latch starts to closing with hook with the help of upward movement of old loop.

4. *Knock Over*
As the head of the needle runs down below the top of the trick, the old loop slides off the needle and the new loop is drawn through it. This process is known as knock over.

5. *Loop Pulling*
Loop is the point when old loop slides from the head of needle and that needle descends more and pulls a length of yarn for adjusting the loop length. That pulled length of yarn is approximately double the distance of which needle goes downward below the surface of sinker.

6. *Cam Track or Cam Profile*
This is illustrated in the Fig. 14 cam track where needle butt moves for the loop formation.

18 The Jack

The jack is the secondary knitting element which is used for the additional designing operation of the latch needles. It is placed below the latch needle. The jack is placed in the same trick used to accommodate the needle butt. Jack also has its own cam system. Jack is commonly used in the superimposed cylinder type and with latch needles which have two ends. These needles and cylinders are mostly used in purl knitting machines. Jacks are the actuating element used with the cams for the specific needles transferred from one cylinder to the other [13].

19 Bearded Needle

The bearded needle found to be the first type of needle to be introduced in the history of knitting. Spring bearded needle comprises of single metal piece. These needles are utilized in the fine gauges up to 60 needles per inch. Beard needle remained in the prioritize usage for four centuries. Bearded needles required an extra attachment called pressor. Bearded needles are commonly used in warp knitting machines.

There are mainly five parts of bearded needle:

- *Stem:* This is the major part of needle which support the loop formation.
- *Head:* This is the continuity of stem at which the needle bent to form hook.
- *Beard:* This is continuation of the hook and flexible part which is pushed toward stem to trapped new loop inside.
- *Eye:* This is also called grove pierced in the stem for receiving of tip of the beard when it is pressed for enclosing the loop.
- *Shank*: This is the bottom part of needle which is attached with the machine part.

Beard needle required an extra element for closing the beard hook.

20 Compound Needles

Compound needles were initially introduced in tricot warp knitting machine in 1946 owing to its higher production rate than warp knitting machine with bearded needles. Warp knitting machines with compound needles used to produce 1000 courses per minute which were far higher than the rest of machines at that time. It is also very much beneficial than the latch and bearded needles in terms of avoiding the distortion of latch [14]. Apart from this in case of compound needle, there is no need to raise so high for the clearing which is another potential benefit. Now with the development of latest electronical and complex machines with diverse patterns and design technology, the efficiency of knitting machines becomes even more higher. Compound needles tend to be stronger than the conventional latch needles, even four times than the strength of latch needles as found in the literature. Compound needles are comparatively less used than latch needle. Compound needles are used in warp knitting machines for the development of advanced functional structures such as hybrid, multiaxial, and composite structures. Compound needles are used to produce fully fashioned or whole garment knitting machine. Compound needles have found their common use in tricot warp knitting machine for the production of sportswear and under wear fabrics. Parts of the compound needle are explained below.

1. Hook
Hook is the part same as present in latch and bearded needles. It also works in retaining of yarn and loop in loop formation process.

2. Tongue/ Piston
Tongue is the part of needle which works like a latch. It closes the hook.

3. Stem
Stem is not solid in the case of compound needle as it appears in the latch and bearded needle. Compound needles have hollow stem in that space tongue moves up and down according to the loop formation action.

For making loop formation in weft knitting, needles are used. There is a question that how needles move for making loop formation? Basically, a path is given to the needles which is especially designed according to the loop which is required to be made. That path is mostly called groove. That groove is made by the joining of cam parts with each other. Needle butt moves in the groove and according to that groove of cam it produces yarn loop. These groves are known as cam profiles and cam tracks.

Needle moves differently in loop forming mechanism of different loops. Cam profiles are also different from machine to machine and needle remains in different position.

21 Dial

In a circular knitting machine, dial is available in the second needle bed which is in circular shape. Dial is positioned on the top of the cylinder and dial has same diameter as of the cylinder. For the accommodation of needles in the dial, it has grooves or peripheral grooves present in which the needles are placed. The upper portion of dial groove extends below to the inner side of the dial. The depth of groove is relative to the needles to prevent the needles entanglement with the yarns. Needle grooves are also called slots which extend toward inner side vertically [14, 15].

22 Cylinder

Knitting machines have cylindrical shaped needle bed in the circular shape made of metal. It has slots for the needle's accommodation. The needles are placed vertically in the cylinders. Needle slots are engraved around the circumference of the cylinder (Fig. 15). The needles move in the needle slots during the knitting cycle. Knitting machines which have single bed or double bed have cylinder. Some double bed knitting machines which perform purl stitches have double cylinder in them.

23 Sinkers

Sinker is the most important element of knitting machine after needles, (as shown in Fig. 16) Needles being top of the list. Sinker plays vital role in stitch formation as the name sinker suggests that the function. Sinkers are positioned perpendicular to the needles on the hook side of the needles and between the adjacent needles. The categorization of the sinkers is based on their functions and performing mechanism which are elaborated below.

Fig. 15 Needles arranged in cylinder tricks

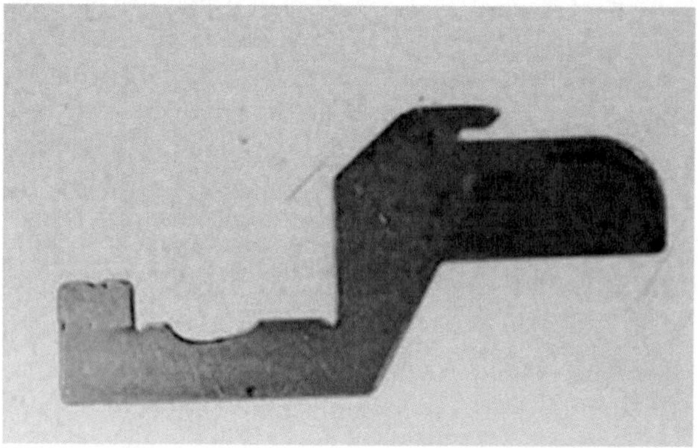

Fig. 16 Sinker

Holding Down Sinkers

In the modern knitting machine, there is another function of sinker to hold down the old loops at a slightly lower level and prevent the old loop from lifting up as the needle rise for the loop clearing. Holding down sinkers produce tighter/compact structures having better appearance. Holding down sinkers are mostly used in double bed knitting machines.

Knock over Sinker

Knock over sinker perform the function of knocking over during the stitch formation. During the course of its action, belly of sinker provides support to old loop while the new loop is drawn through it. In different knitting machines such as warp knitting (Tricot and Raschel), the sinkers are in specialized shape for performing the functions.

Sinkers Ring
Sinkers ring consists of the sinkers cams which are arranged adjacent to each other. Each cam on the sinker ring undergoes the sinker action of advancement and retracting [16].

24 Designing Elements of Knitting Machines

It is very well known that different process and treatments play important roles in the designing of fabrics. From fiber to fabric at each stage, the appearance and aesthetics of textile products can be decided. The knitting machine also plays an important role in designing the fabric with the different knitting elements on machines. Besides the basic assembly of a knitting machines, extra attachments are introduced in the knitting machine to offer a wide range of design versatility.

25 Cam Plate Mechanism

Individual needle selection technique of designing provides a wide variety of designs having single or multiple color. The basis of individual needle selection is the positions of needle hook during the clearing action. Needle may reach to the one of three positions which are clearing height, tuck height, and floating height. Latch needles are preferably and mostly utilized in this technique. Some techniques and elements are used in order to perform individual needle selection and produce variety of designs. The techniques of individual needle selection are elaborated further.

Some rules are there according to which the designing technique of individual needle selection is applicable and has its effect on the designs.

- If there is fixed cam track and each element have butt of same length, then it will pass through the identical path and produce similar stitch in relative wale of that feed.
- Different types of stiches can be produced in the same wale if the butt follows different path with the support of successive cams.
- Different types of stitches can be produced in the adjacent wales if butts of adjacent elements pass through different paths.
- Similar stiches in each wale of every feeder course can be produced in the case of same arrangement of fixed cams.
- The wales wise width of the design is determined by the range of independently selected needles means the number of needles which are individually selected.

The designing technique with the support of *different butt positions* on the needle stem can be endured in the machines with multiple cam tracks (Fig. 17). Multiple cam system is the basic principle of designing with different butt positions (Fig. 18). It is applicable on the machines with multiple cam track such as single jersey

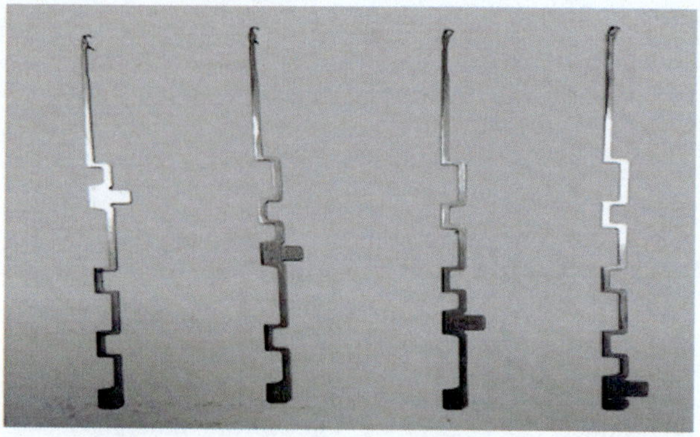

Fig. 17 Needles with different butt positions

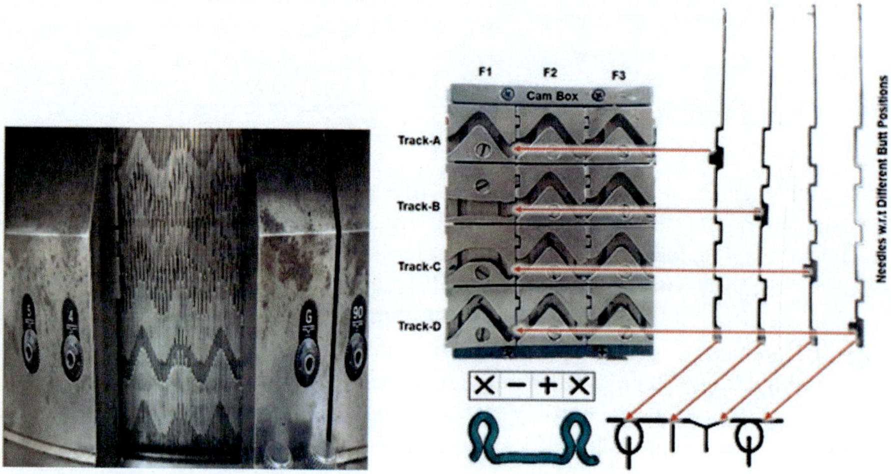

Fig. 18 Multicam tracks

machines with multiple cam track in which the needle butts placed in between 2 and 5 cam track. Theses cam tracks have exchangeable knit, tuck, and miss cams which are fixed. It is also applicable in the interlock machines in which mostly two cam tracks are used. It is known that the stich produced by a needle is determined by the lifting height a needle attain in its trick. For proper understanding of the effect of different butt positions of needles on the knitting designs, let us have a scenario in which all the needles' butts are at same position on the stem of needle. In this case, the needle will pass through the same cam path and result in same type of stiches produce. This will ultimately produce plain fabric. So, a wide range of designs can be produced with the help of needles with different butt positions. Different butt positions will decide the lifting height of the needle in the raising cam either the needle will follow the path of knit stich, tuck or miss stitch. Butt positions at different heights have direct impact on the width of the design.

26 Jacquard Knitting

Jacquard mechanism involves single needle control of knitting machine. Each needle from any point of knitting bed or cylinder can be selected for specific stitch. Jacquard offers design area of unlimited depth and width depending on the number of needle selection. Jacquard is used for the designing of complex patterns. It is generally used to produce elaborative designs. Both flat as well as circular knitting machines are capable of jacquard designing. Jacquard is further divided into subcategories based on the number of needle selection and control. Those classifications are: Intermediate jacquard, which has the capacity of controlling up to 24 wales; medium jacquard, which has the capacity up to 48 wales; and full jacquard, which can control up to 144 Wales.

According to the principle of needle selection, jacquards may be of either mechanical or electrical (electronic) type, which are explained further.

Mechanical

Mechanical jacquards use the design elements to create required pattern. It can be by the punched cardboard same as the weaving or needles discs mechanism, and pattern wheel. Pattern wheel is shown in Fig. 19.

In punch card mechanism, the selectors are sorted in the groups of 48 which are called automat and these automats are controlled by the punched card. Automats are arranged around the circumference of cylinder. Jack is placed below each needle, the end of jack is supported by the pivoted lever which perform the action of lifting called lifter lever. There is another lever called automat lever which is pivoted and rest on the outer end of lifter lever. The outer end of automat lever holds a pin which is spring loaded and positioned on the top of punch card as it passed over the grooved

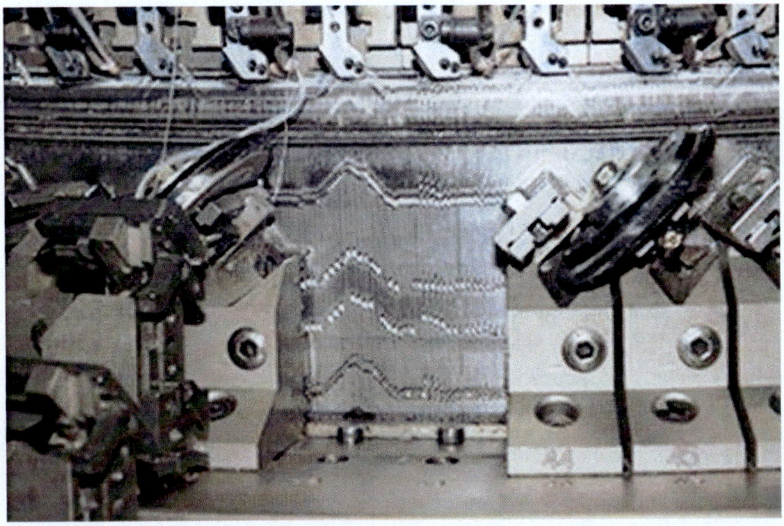

Fig. 19 Pattern wheel

roller. The roller is driven by the machine and turned intermittently. Punched cards force the corresponding needle to produce a normal knit loop. When there comes hole in the punch card, the automat passes through the hole and reached onto the groove of roller. The roller turns upon the selection and tends to move forward the punch card. This causes automat lever to move forward as the pins of automat lever are entered in the holes. In case when there is no hole punched, the respective pin rests on the film and respective automat lever doesn't move forward. The action of cam is not transmitted to the lifting lever as a result needle remain in the miss position no knit formation.

Pattern wheel is a simple device work as designing element in knitting machines. It is a small device which utilizes least space and performs its function. Pattern wheel is an economical and cheap designing technique. Pattern wheels have separate raising cams which are available in the form of pattern bits for the selection and movement of individual elements even up to three different positions inside the tricks. Jack, which is composed of metal plate, is placed on each side of pattern wheel. The jacks have rectangular cross section in with the outer end portion is twisted relative to the face of jack plate which is the same angle which is formed with the peripheral teeth. Peripheral teeth are on the wheel [17]. Pattern wheels form a peripheral tooth just like helical gear tooth. The selection area for designing purpose follows the spiral path of the courses produced by the feeders around the fabric tube/tubular fabric. Pattern wheel is mostly used in single jersey machines. Pattern wheel can be placed either in inclined position for the selection of needle or in horizontal position for the selection of plush sinker. The designing pattern utilizes bits which are inserted into the tricks or in some cases break-off teeth on discs which are already prepared. The pattern wheels are introduced into the same gauge as of the cylinder needles and driven in the opposite direction. This drive can be given due to the needle butts encountering the tricks or by the gear of cylinder. The design produce by the pattern wheel is in spiral form. Inclined pattern wheel is placed at the angle of 20–40°.

Electrical

Designing technique involving electromagnetic needle selection is called electronic or electrical jacquard. In this technique, an electronic impulse is generated which is used to provide potential to an electromagnet. This electromagnet performs the selection of the needle and determines its position during the action of loop formation. This electromagnet generates minor motion which is magnified by the help of mechanical motion of other elements. The input to this electro-magnet can be given in different forms such as CAD drive as well which is the latest development. In first-generation electric jacquard, photo-films were utilized for providing sufficient input. In electronic jacquard, jacks or sliders are positioned below the needles. As the clearing cams are not present in the needle cam track, the clearing function is performed by these jacks. These jacks have butts which are either controlled by the jack cam track determined by its position in the trick or beyond its control. The two positions of jack are controlled with the help of two flat springs. One spring, which is known as retaining spring and is placed at the back of jack, tends to push the jack outside so that butt come out of the jack under the control of jack cam track. For the proper accommodation of retaining spring, the part of trick is deeper as compared to the rest of trick. The other spring is known as control spring which is placed in front of the jack. This control spring in case of not attracted by the electro-magnet or solenoid causes to push the jack inside the trick against the retaining spring. In case control spring is attracted by the electro-magnet, retaining spring pushes forward the jack. In different scenarios different inputs which may be Photo-film, magnetic film, CD, floppy EPROM, the design pattern is scanned by appropriate device. This reading is done course wise, and the respective signal is sent to the electro-magnet. The electro-magnet gets energized or not depending on the nature of signal. If electro-magnet is energized, it will attract the control spring so that it will bring the jack under its jurisdiction to the control of jack cam track. The jack raises up with the help of cam track of jack, and this results in needle to move up to the clearing height which ultimately results in loop formation. If in the case of no potential given to the electro-magnet by the input signal, the control spring tends to push the jack inwards to the trick; as a result, jack is not lifted up by the jack cam track. This will ultimately lead to staying of needle in the stationary position in lower position which results in formation of float or miss stitch. In the latest development, elements having piezoelectric characteristics are utilized for the better performance and selection of needles.

The double jersey machine model V-LEC3DGTY2 presented at ITMA was a computerized controlled machine capable of stitch transfer from cylinder to dial and dial to cylinder. This stitch transfer mechanism can be done on all feeds. The machine has the capacity to knit garment length with separation of welt with the help of a draw thread [4]. The mini-jack circular knit machines are most suitable to produce jacquard fabrics. Some manufacturers offer jacquard technology in single jersey machines.

27 Intarsia

Intarsia is a designing technique to produce different colored effect on the knitted fabrics. It is done using yarn of different colors which produce a specific-colored fields/blocks of adjacent courses. So, intarsia techniques produce blocks of different colors and designs on the knit fabrics. It allows the occurrence of particular-colored loops on the different sides at desired intervals. Intarsia technique is preferably used to produce geometrical designs instead of figured design covering minor area. Intarsia is comparatively a slow process. MBI technology (MBI Technology Co., Ltd, South Korea) is a widely acknowledged manufacturer of machines having intarsia designing technology. Usually, intarsia technique is performable on the flat knitting machines, but MBI technologies offers intarsia technique in circular knitting machines. This technology offers 42 different colored presentation in single course in the different areas of knitting body. The first model of circular machine having intarsia technology was S1342. Some others knitting machinery manufacturers developed the flat knitting machines with intarsia technique some of them are SHIMA SEIKI Japan, STEIGER Switzerland, STOLL Germany, and PROTTI Italy.

- Flat knitting machine may be either jacquard or non-jacquard
- It may be single needle bed (for plain knit) or double needle bed (opposite flat bed, V bed)
- Fully fashioned machines

28 Flat Knitting Machine

Flat knitting technology offers the diversity in structures up to the three-dimensional structure. The limitations are in terms of productivity as the discontinuous carriage motion and limited numbers of feed system are 4 to 6. Circular knitting machines offers high productivity owing to the higher number of feeders up to 144. To produce 3D knitted fabrics on the flat knitting machine, the commonly known term is needle parking. This technique includes two operations, which are extension in which the number of active needles increased and constriction in which the number of active needles decreased. During the process of active and inactive of needles, some needles remain in active for the certain time during which they hold the old loop last knitted in the cycle.

29 Types of Flatbed Knitting Machines

Flatbed machines are available in different types which are explained following section:

Classification w.r.t. Drive

Manual Flatbed Machines
Manual flatbed machines are hand-driven machines, So for the mentioned purpose (Hand-drive) a handle is available in carriage. The operator uses the handle for the traverse motion of carriage from one side of machine to other.

Automatic/Power
In automatic flat knit machine, carriage is attached with the motor to provide traverse motion. In these machines as the carriage reaches from one side to the other, two-way switch tends to change the current direction in the motor and as a result carriage moves in opposite directions.

Classification w.r.t. Bed

Single Bed Flat Knit Machines

These types of flat knit machines are used for knitting single jersey or plain fabrics. The single bed flat knitting machine with intarsia is used for sweater production. This machine comprises yarn feeders, a yarn carrier, and single bed. The machine has the capacity to accommodate 3, 5, 8, 10, and 12 needles per inch of the bed. This latest technology of flat knitting machine results in reduction of labor cost up to 10%; it was also found that reduction in material costs was about 6.2%. The efficiency of machine also increases by lowering the production time by 11.5%. This machine has logarithm for the optimization of production [21].

Double Bed Flat Knitting Machines

These are used for the development of rib and purl knit structures. Double bed machine has the capability to convert it to single bed machines by easily detaching one bed. In double bed flat knitting machine, the interlock gating is not possible. Same as the circular knit machines, the flat bed purl machines utilize single set of double hooked latch needles. The single set of needles simultaneously knit and transfer stich on both beds.

Main Parts of Flat Knitting Machine

Main parts of flatbed knitting machine are quite similar as of the circular knitting machine.

30 Needle Bed

Needle bed is a metallic plate which has slots/tricks embedded in it for the accommodation of needles. Needles move to and fro in these slots during the loop formation cycle. These slots should be smooth enough to facilitate the easy movement of needles. It is noteworthy that the slots of one bed should be in between the tricks/ slots of other bed rather than positioned exactly opposite to each other [18]. There are two types of flat knitting machines available, i.e.,

1. V-Bed Rib M/C
2. Flatbed purl or Links-Links M/C (Fig. 20)

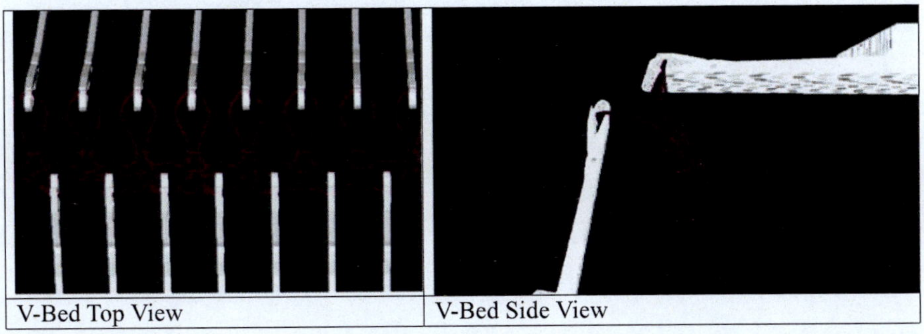

Fig. 20 Flat knitting

In V-bed flat, the two beds are positioned at an angle of 90–105° to each other. Flatbed purl machines have two horizontal needle bed. They are mainly used for cable stitch, basket purl, and lace patterning. The needle bed facing toward operator is called front bed while the other bed is called rear bed. The length of needle bed in manual machines is about 120 cm while in automatic machines length is up to 320 cm.

The relative lateral moving of beds is called racking which is done for the designing purposes. For example, 3D cable stitch is an example of design produced using racking principle.

31 Carriage

Carriage plays a vital role in loop formation process in flat bed knitting machines (Fig. 21). It controls the needles by cams. Plates include "cam locks" bearing the drive and control system of needles, i.e., cams. To produce various types of stitches, needles must follow the path offered by the cam tracks. Cam boxes provide a platform for the cams. Usually there are three types of cams are placed in Cam box which are Raising cams, Lowering cams and Fixed Cams. The Raising cams & Lowering cams during loop formation process these types of cams tends to lower the needle butts for knocking over and Fixed cams which support the smooth shape of the cam track. Carriage has double function: (1) to select the needle and make them raise or lower to form the stitch, and (2) to select and drive the thread guide which feed the needles.

Carriage is identical on both sides, so no problem in operating from right to left or vice versa. For the proper running of carriage, guide rails are positioned on the lower side of needle bed. In some automatic flat knitting machines for the higher production rate, double cam system is used in which two sets of cams are placed side by side. Both sets are supplied with the yarn from the separate yarn carrier. The mentioned arrangement of double cam system results in the production of two courses in a single travers motion of carriage. This double cam system results in increasing the length and weight of carriage.

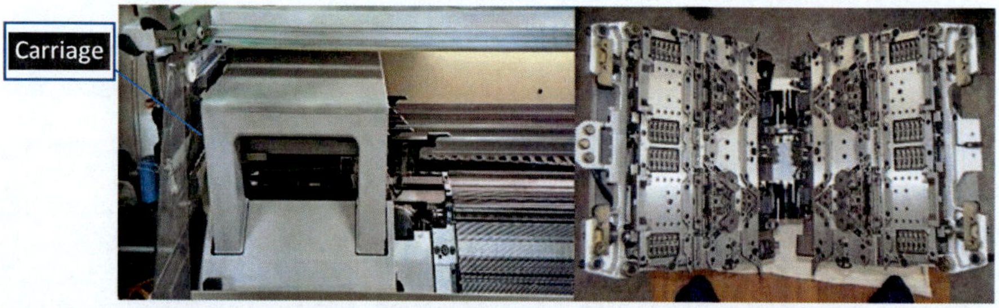

Fig. 21 Carriage assembly of flat knit machine

32 Yarn Feeding

Yarn bobbins or cones are set on a table connected to rear frame of the machine (spool rack). Yarn is fed to the needles from the yarn packages using the attachment called yarn carrier. Yarn is supplied separately to each yarn carrier. Yarn feed system of flatbed knitting machine is shown in Figs. 22 and 23. Yarn carrier is attached with the assembly which moves along the length of needle bed supplying yarn to the needles. Yarn carriers are selected by the carriage and moved along the needle bed length as per the design requirement. Different types of tensioners are also installed on the machine to maintain the tension on the yarns. The yarn is fed through assembly to yarn guide of carriage.

H. Stoll GmbH & company is one of the leading manufacturers of knitting machines. Produced new model of flat knitting machine CMS ADF-3. The machine has superior yarn feeding technology which offers great versatility in structural technique and color combination. Machine has 16 rails having two yarn carriers accommodated in each rail. Total 32 yarn carriers are powered by a motor and have

Fig. 22 Yarn feed system of flat bed

Fig. 23 Yarn feeder of flat bed knit machine

the capability to be positioned vertical or horizontal. Machine is also equipped with intarsia technology and able to produce color feed which is less than one inch [19].

33 Fabric Takedown

Fabric takedown mechanism plays a vital role in loop-clearing process and loop formation. For the proper loop formation loops, formed loops and resultant fabric should remain in intact in their place. For the mentioned purpose, optimum tension on fabric formed is required. To hold the fabric down and intact, weight is applied. In latest machine, the fabric takedown is done by the takedown rollers which applied tension on the loops and wrap the cloth thus produced around themselves. The rollers supported takedown mechanism result in the application of uniform tension and continuous knitting process.

34 Monitoring System of Flat Knit Machines

Monitoring system of the knitting machine is very important for the better efficiency and productivity of the machine.

Fouda et al. presented a monitoring system of the flat bed knitting machine. This monitoring system senses the various forces that occur on the different parts of machine during knitting action. A sensing element is placed on the carriage of flat knit machine in which during the motion of needles stress is applied on the sensing element. It was found that different parameters affect the force applied on the needles. For example, increase in loop length results in an increment in the applied force. It was also found that knitting force decreases as the yarn ply twist increases. From this type of monitoring system, the optimum machine setting can be decided for the better efficiency of the machine [20]. The technical features of the modern developed electronic V-bed flat knitting machines which give them the capability of manufacturing complex-shaped engineering structures are: The CAD-patterning, Loop transfer, Individual needle selection capability. The usage of intarsia yarn feeders together with the presence of holding down sinkers are the epic features.

35 Fully Fashioned Seamless Flat Knit Machine

Fully fashioned knitting technology eliminates the resource consumptions of cutting and sewing. In this knitting technology, the different parts of knit garment are produced separately. In 1995, Shima Seiki introduced fully fashioned garment technology which is continuously improved with the passage of time. Eliminating the cut and sew process allows for "quick-response production" (QRP). In seamless technology, shaping of certain parts can be done or whole garment can be possible.

Shima Seiki presented its latest development in fully fashioned computerized flat knit machine in *Techtexil North America* exhibition in Atlanta. The machine model is SRY123LP which has better control on the press down of individual loops with the help of loop presser beds which is placed on the top of conventional needle bed. The machine presents better designing capabilities in terms of partially knit patterns and inlay patterns. Another model (SWG061N2) of fully fashioned flat knit machine by the same manufacturer can produce seamless 3D products. The same model has the capability to produce small, knitted tubular products [22].

36 Glove Knitting Machine

Gloves can be produced by cut and sewn method or by glove machine for seamless gloves as shown in Fig. 24.

37 Latest Advancements in Knitting Machines

For the design versatility and higher efficiency, the computerized machines are the best tool. Computer-aided designs can be made using specialized applications. Special textures, color combinations, and 3D designs can be easily developed using computerized knitting machines. These designs can be made by manipulating the loops between the needle beds. Vidya Naryanan et al. presented a 3D visual interface for the development of complex designs using augmented mesh stitches [23]. Seamless garments are the latest technology involved in the production of three-dimensional whole garment on knitting machine. It is perceived that this latest technology evolved from the hosiery knitting. Eliminating the cutting and sewing process yarn is directly formulated into garment.

Santoni, an Italian based manufacturer, is leading organization in providing advanced and automatic seamless knitting machines. Seamless technology is available in both circular and flat bed knitting machines. In circular knitting machines, the selection of optimum diameter of cylinder depends on the choice of finished garment size. Seamless technology proves to be very helpful in reduction of waste

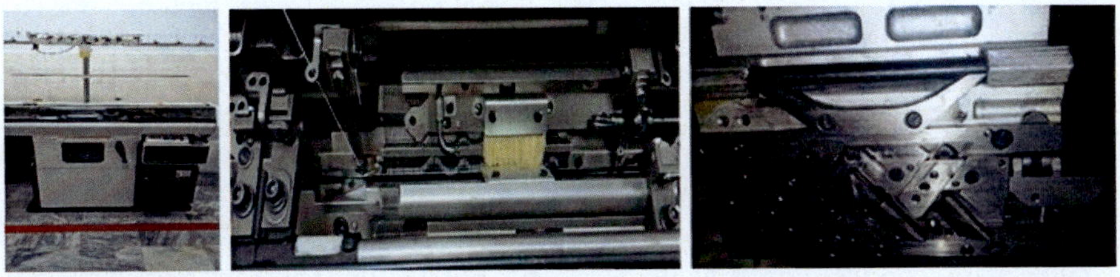

Fig. 24 Flat bed glove knitting machine carriage

and process cycle. Seamless technology offers better comfort attributes in dresses by eliminating the seams and stitches. Different designing software are used for the development of seamless knitted garment. Digraph and Pulsar design software by Dinema and other designing software from Santoni 3D technology are very useful. Shima Seiki is considered to be the first to introduce fully fashioned/seamless whole garment knitting technology.

Takedown mechanism is a very important parameter in this concern deciding the accuracy in the whole garment technology, as during the development of seamless garment on the knitting machine, an incremental tension is required to keep the different parts of garment on their optimum position. By eliminating the seams, different properties of garment become much better such as the drape property enhanced as the seams who stops the flow are no longer exist now. This technology basically works on the digital designing and computerized software. Computerized flat knit machines are the best choice for the adaptation of this technology. The production of garment starts from the bottom. Three tubes are formed, wide tube for body portion and two small tubes for sleeves. First of all, three welts are formed and then simultaneously the production of three tubes and at the end three tubes are joined at under arms, neck, and shoulder portion. Transfer stitch between two beds is the base behind this technique [24].

Latest development in the field of seamless garment is slide needle design technology. Contra knit technology explained earlier also plays a vital role in the development of whole garment. Slider is flexible, comprising two parts which extended up to the needle hook have great potential for better in complex transfer. Slider mechanism facilitates the transfer mechanism efficiently and permits the needles to be positioned in the center by eliminating the transfer clip. This helps in the achievement of symmetrical loop formation. Commonly six techniques are used for the seamless technology: Knit, Tuck, Miss, Receive, Racking, and Transfer. In case of slider technique, it offers 12 techniques, the six techniques other than traditional are Split stitch lateral transfer, second stitch, lateral transfer, split front and back, holding and inlay which provide great diversity in the knitting in pockets, button holes, collars and many more trimmings. The slider mechanism tends to increase the variation in stitch characteristics, offering a great range of gauges to be utilized for the seamless technology for the production single garment. It is noteworthy that seamless technology is extending to the field of automotives seamless seat covering and in medical field to produce the different shaped scaffold for medical purposes. Shema Seiki launched six different models of knitting machine having seamless fully fashioned garment technology [25].

Santoni Single jersey circular machines Model SM8-TOP2V electronic machine offers whole garment technology. The machine has 8 feed with 2 points of selection needle by needle per feed. The whole garments provide their applications in under wear, sportwear, outerwear, and sanitary wear. The machine also facilitates the production of Terry [26]. Another model SM4-TL2 of single jersey machine with whole garment technology by Santoni is available in the market. The machine has 4 feed system with reciprocal movement. There are 2 selection points available per feed and per rotation [27]. SM8-TOP4J is single jersey electronic machine with diameter of 12inches to 22 inches [27]. Seamless technology is also offered in Double jersey circular machine Model SM-DJ2T. Machine has the rib knitting capability with 2 selection points on dial as well as cylinder. Machine is available in wide range of Gauges €13, 14, 15, 16, 18, and 20 needles per inch. Pneumatic and mechanic take-down mechanism is installed. Dial height and cylinder rotation can be adjusted [28].

Flat knitting technology offers greater design versatility and greater flexibility than circular and warp knitting technology. Hong Hu et al. used two computerized flat knitting machines, model Stoll CMS 530 E3.5.2 and Stoll CMS822 E 7.2, for the production of auxetic fabrics. The auxetic fabrics have negative poison ration, so these fabrics expand as they are stretched. Three different structures were utilized for the production of these fabrics: rotating rectangle, reentrant hexagon, and foldable structure [29].

The jacquard is an advanced designing technology available in both circular and flatbed knitting machines. As mentioned earlier, jacquards are of two types: mechanical and electrical.

References

1. R. Rutt, *A History of Hand Knitting* (Interweave Press, 1987)
2. China knitting machine manufacturer. [Online]. Available: China Knitting Machine Manufacturer, Circular Knitting Machine, Computerized Knitting Machine Supplier – Quanzhou Baiyuan Machinery Science & Technology Co., Ltd. (made-in-china.com). https://baiyuan.en.made-in-china.com
3. Rib Circular Knitting Machine, Brochure, p. 72. Available online: Address: Xinda Knitting Machine, Wuli Industrial Park, Jinjiang, Fujian, China
4. A. Richard Horrock, C. Subhash Annand, *Handbook of Technical Textiles*, Vol. 2, 2nd edn. (Woodhead Publishing Series in Textiles, 2003), pp. 1–5.
5. D. Semnani, Chapter 7: Advances in circular knitting, in *Advances in Knitting Technology*, (Woodhead Publishing Limited, 2011), pp. 171–192
6. S. Loeb, United States Patent (19) U. S. Patent Oct 13, 1987. Foreign Appl. Prior. Data, no. 19 (1987)
7. Y. Onashi, M. Enishi, A yarn feeding device for circular knitting machine. Patent Application Number EP1939340 (A1) – Precision Fukuhara works. Ltd. 02-07-2008
8. Fabric take-up mechanism for circular knitting machine. European Patent Application Number 93116325 (1993, October 8)
9. A.P. Catarino, A.M. Rocha, J.L. Monteiro, F. Soares, A system for knitting process monitoring and fault detection on weft circular knitting machines. *Semantic Scholar* (2004, June 22)
10. Advanced fabric inspection system for circular knitting machines. Available online: https://www.knittingindustry.com/circular-knitting/advanced-fabric-inspection-system-for-circular-knitting-machines/
11. A. Catarino, A. Rocha, J.L. Monteiro, F. Soares, Knitting process surveillance using time and frequency analysis. IEEE Int. Symp. Ind. Electron. **IV**, 1563–1568 (2005). https://doi.org/10.1109/ISIE.2005.1529165
12. A. Catarino, A. Rocha, J. Monteiro, Low cost sensor for the measurement of yarn input tension on knitting machines. IEEE Int. Symp. Ind. Electron. **II**, 891–896 (2003). https://doi.org/10.1109/ISIE.2003.1267939
13. D.J. Spencer, *Knitting Technology: A Comprehensive Handbook and Practical Guide*, 3rd edn. (Woodhead publishing series in textiles, 2001, April 27).

14. V. Kothari, Compound needle warp knitting machine. Knitt. Views December, 24–25 (2012). [Online]. Available: www.vasantkothari.com
15. I.W. Grothey, *Dial for Knitting Machines*. United States Patent Number 2027276. Filed 2 June 1934
16. Circular knitting machine with sinker cams facilitating high speed operation. International publication number WO2010/077817A3, Monarch Knitting machinery corp. 14 December 2009
17. *Knitting Machine Pattern Wheel*. United states patent number US3857259.
18. S.C. Ray, Flat bed knitting. Fundam. Adv. Knitt. Technol., 101–116 (2012). https://doi.org/10.1533/9780857095558.101
19. Stoll launches new flat-knitting machine – Advanced Textiles Source. *Textile Technology Source*. Available online: https://advancedtextilessource.com/2013/10/15/...
20. A. Fouda, A. El-Hadidy, A. El-Deeb, Knitting force measurement on flat knitting machines. J. Text. **2014**, 1–9 (2014). https://doi.org/10.1155/2014/546472
21. S.M. Huh, W.J. Kim, Productivity optimization for intarsia single bed flat knitting machine using genetic algorithm. J. Text. Inst. **113**(1), 33–44 (2022)
22. *Advanced Flat Knitting Machine for Technical Textiles to Be Shown in Atlanta*. [Online]. Available: https://www.innovationintextiles.com/advanced-flat-knitting-machine-for-technical-textiles-to-be-shown-in-atlanta/
23. V. Narayanan, K. Wu, C. Yuksel, J. McCann, Visual knitting machine programming. ACM Trans. Graph. **38**(4) (2019). https://doi.org/10.1145/3306346.3322995
24. F. Lau, W. Yu, Seamless knitting of intimate apparel, in *Advances in Womens Intimate Apparel Technology*, 1st edn. (Woodhead Publishing series in Textiles, 2016). pp. 55–68. https://doi.org/10.1016/B978-1-78242-369-0.00004-9
25. Seamless Knitwear_ New Technology Ensures One-Piece Construction with Minimal Wastage – Apparel Resources Bangladesh. Available online: https://in.apparelresources.com/technology-news/knitting-technology/seamless-knitwear/
26. SM8-TOP2V – Santoni. Available online: https://santoni.com/en/products/seamless/sm8-top2v
27. B.K. Anderson, D. Ph, *Seamless Technology*, pp. 1–6. Brochure available online at: www.santoni.com
28. B.K. Anderson, D. Ph, *Seamless Technology Double Jersey Santoni*. Brochure available online at: www.santoni.com.
29. H. Hu, Z. Wang, S. Liu, Development of auxetic fabrics using flat knitting technology. Text. Res. J. **81**(14), 1493–1502 (2011). https://doi.org/10.1177/0040517511404594

Weft-Knitted Structure and Their Effect on Fabric Properties

Adeel Abbas, Hafsa Jamshaid, and Rajesh Mishra

1 Introduction

Knitting, considered to be bending straight yarn into loops and then intermeshing those loops for fabric formation, sounds simple. However, different loop geometries configure specific textures on the fabric surface due to unique constructions as shown in Fig. 1. Such geometries are responsible for inducing the required physical, mechanical, and comfort properties in knitted fabrics. Weft knitting architecting fabrics in course-wise direction comprise additional classifications of single jersey, rib, and interlock-knitted fabrics. In each type of weft-knitted fabric, the loop geometry alteration principle is followed, creating knitted structures with specific names, notations, and properties. Merely, a weft-knitted fabric's cover factor can be altered by different stitches, and properties dependent on the cover factor are eventually changed.

Such architectures are engineered through three basic weft knitting loop types, termed knit, tuck, and miss /float. The appearance of these stitches, effect on fabric properties, and formation process are different from each other. They are also incorporated due to their distinctive and effective appearance as shown in Fig. 2. Knit stitch is the basic stitch that works as a constructing unit of knitted fabric. Without knit stitch, fabric cannot be produced. This stitch has a major role because it is used to connect adjacent loops. In knit stitch formation, the needle completely follows the cam track to both the highest and lowest points. Knit loop is formed by

A. Abbas (✉)
Faculty of Textile Engineering, National Textile University, Faisalabad, Pakistan

H. Jamshaid
School of Engineering and Technology, National Textile University, Faisalabad, Pakistan

R. Mishra
Faculty of Engineering, Czech University of Life Sciences Prague, Prague, Czech Republic

H. Jamshaid, R. Mishra (eds.), *Knitting Science, Technology, Process and Materials*, Textile Science and Clothing Technology,
https://doi.org/10.1007/978-3-031-44927-7_4

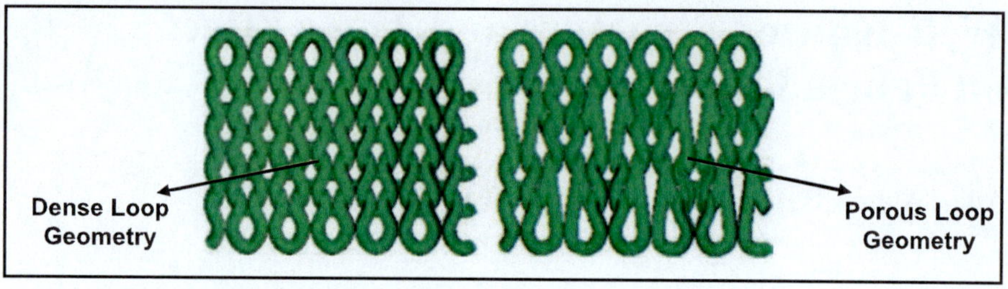

Fig. 1 Knitting architectures

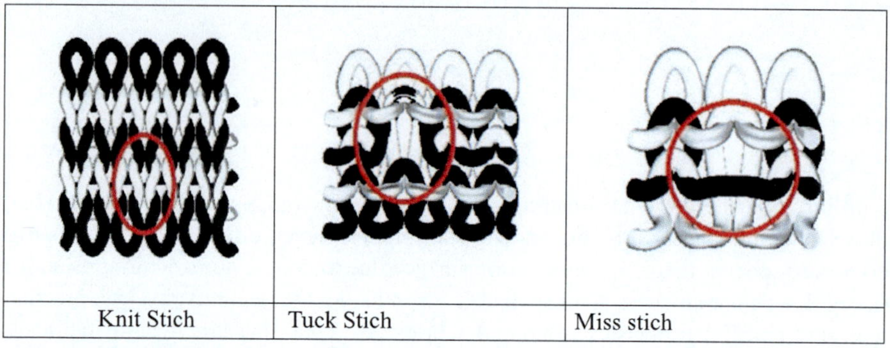

Fig. 2 Appearance of stitches

providing a knitting needle with complete clearing and feeding height [1]. Old loop is cleared into the needle stem, and the new loop's yarn comes into the needle hook. But in the construction of other stitches, the needle does not complete a full knitting cycle. Tuck stitch is formed when the needle without knocking over the old loop is fed with the new loop, and then the old loop is knocked over with the newly fed loop. For tuck stitch, the needle only takes feeding height, the old loop is not cleared, and the needle's hook keeps two yarns simultaneously in its hook. Tuck stitch is used for getting a fancy effect of colored yarns. Tuck stitch is used to produce open work effects, improve the surface texture, enable stitch shaping, reinforce, join double-face fabrics, improve ladder resistance, and produce mock fashion marks. Miss or float stitch is also composed of a held loop, one with more floats and knit loops. Float stitch is a straight line of yarn in appearance. Actually, when the needle does not rise and fails to get a new yarn for new loop formation, then a miss stitch is formed. Needles do not take any height for miss loop and move forward keeping the previous loop yarn in the hook. However, the miss loop appears as a float on the backside of the fabric. A floating thread is useful for hiding unwanted colored yarn when producing jacquard designs.

2 Weft-Knitted Structural Notations

2.1 Verbal Notation

It is the simplest design description technique where structural repeat of each course is described verbally in a sentence, e.g., a design consists of first course of all knit stitches and second course with tuck stitches on even needles and knit stitches on odd needles. Similarly, the whole design repeat can be calculated by seeing the structural knitted cell of each course. The method does not provide any visual representation of repeats; hence it can be confusing while describing complex structures [2]. Thus, verbal notations are not preferred for complex structures.

2.2 Graphical Notation

Graphical notation involves drawing the fabric structure loop by loop with hand as it appears physically in fabric swatch, as shown in Fig. 3. The technique requires perfect drawing skills for a person to describe knitted structures. Lack of overall drawing neatness could be confusing. Moreover, graphical representation drawing is hectic and time-consuming, which makes the method least preferable [2].

2.3 Symbolic Notation

The representation technique uses rows and columns for design description. Courses are emphasized as rows and wales as columns. So, the position of each stitch can be identified with reference to specific wale and course number. Each stitch type is demonstrated by a special symbol (Table 1) [2]. The notation method is quite simple

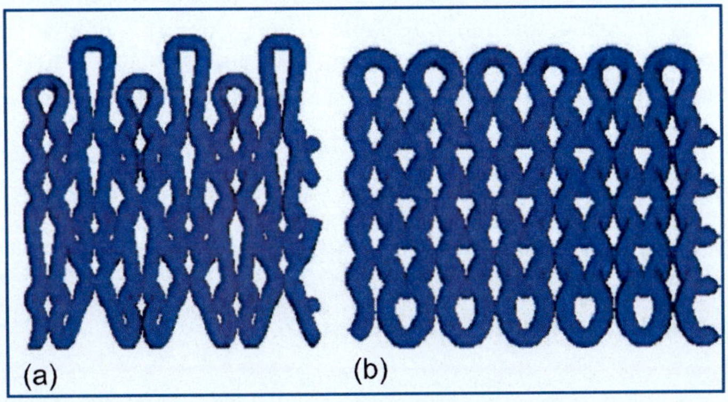

Fig. 3 Graphical notation

Table 1 Symbolic notations

Stitch type	Notation
Knit loop	×
Tuck loop	•/+
Miss loop	–

Table 2 Diagrammatic notations

Stitch types	Cylinder stitch notation	Dial stitch notation
Knit loop		
Tuck loop		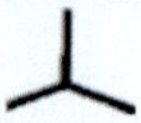
Miss loop		

and easy to understand as compared to verbal and graphical representations. Moreover, drawing of symbolic notations is simple that enables anyone with least drawing skills to draw symbolic notations in an explicable way.

2.4 Diagrammatic Notation

Diagrammatic notation is a widely used knitted fabric architecture representation technique. The method is more convenient to use when describing the patterning mechanism of weft-knitted fabric while relating with cams and needles arrangements. Diagrammatic notation also facilitates a stitch description with reference to its wale and course number, like symbolic notation diagrammatic notation has specified symbols for knit tuck and miss stitches (Table 2). Symbolic notation becomes difficult to understand while discussing double jersey weft-knitted structures. However, diagrammatic notation is feasible to easily describe double jersey structures as shown in Fig. 4. Each stitch is mounted on a needle while drawing diagrammatic notation. Cylinder-knitted stitches are kept upside down, and dial-knitted stitches are kept upward [2].

3 Weft Knitting Patterning Mechanisms

3.1 Conventional Cam Plate System

Knitting machine elements are responsible for variable weft-knitted structures construction. The elements majorly include CAM plates when specifically conventional weft knitting is discussed. CAM plates provide tracks for needle movement and

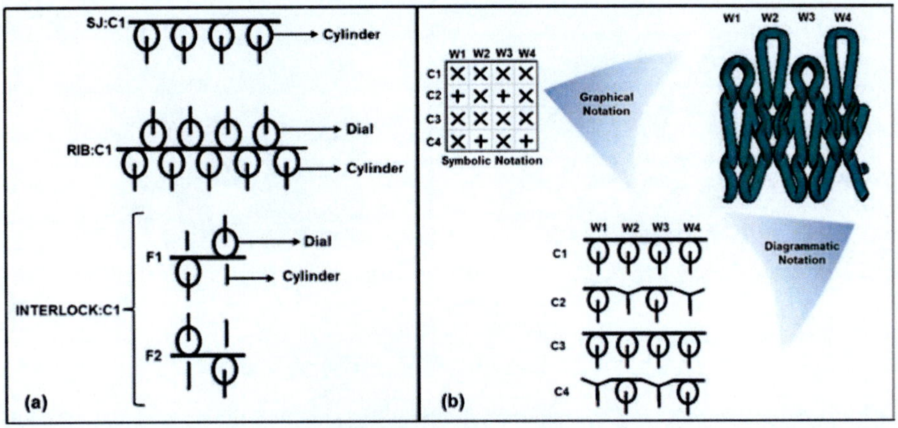

Fig. 4 (**a**) Plain single jersey, rib, and interlock diagrammatic notations (**b**) diagrammatic notations with reference to symbolic and graphical notations

uplifting, which governs the end fabric architecture [2]. Three basic types of CAM plates are used in weft knitting, termed knit, tuck, and miss CAMs.

3.1.1 Knit CAM

Knitting needle follows a designed path where it moves forward in machine cylinder (considering circular weft knitting) followed by upward and downward movements due to CAM tracks [3]. Knit cam provides clearing height to the needle where the old loop is cleared reaching the needle stem while the new loop's yarn is being fed into the needle's hook. Knit cam's path seems like an inverted "V" having an ascending steep path, peak center point, and a descending steep path as shown in Fig. 5. Stitch geometry formed via following knit cam path is termed knit stitch. Knit stitch is the simplest and frequently used weft-knitted stitch. Yarn consumption and stretchability of knit stitch is higher as compared to other basic weft knitting geometries. Hence, mechanical and physiological properties can be altered using novel combinations of knit stitches with some other loop designs. Fabric comprising knit stitches shows greater extensibility than tuck and miss stitches.

3.1.2 Tuck CAM

Tuck cam plate provides the knitting needle a path where it has the same ascending and descending movements, but the upward movement faces a difference. Knit cam provides clearing height to the needle; however, tuck cam takes the needle to only feeding height as shown in Fig. 5. Hence, the needle raised on the tuck cam feeds new yarn into its hook while the previously knitted loop's yarn relays onto the needle's latch surface, while needle stem comprises a second, old, knitted loop (held

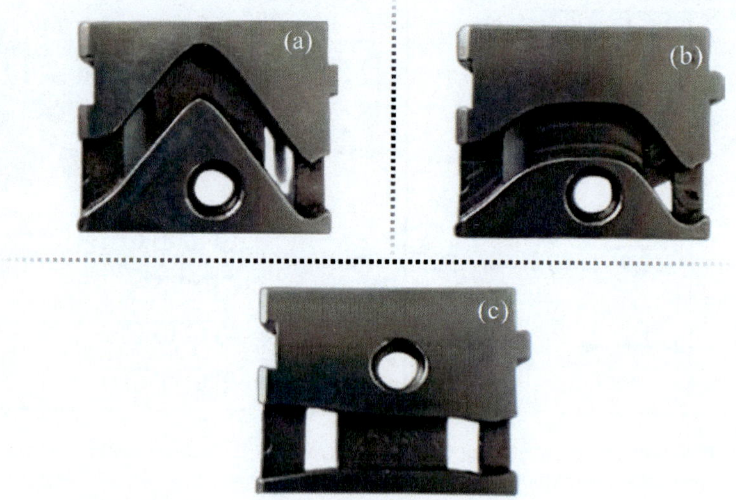

Fig. 5 Knitting CAM plates (**a**) knit (**b**) tuck (**c**) miss

loop), which helps in latch closing during downward movement of the needle. Tuck cam plate's governed stitch geometry is termed tuck stitch. Tuck stitch has the second highest yarn consumption among basic weft knitting geometries. The yarn consumption directly relates to knitting needle height, the more height the needle takes during yarn feeding the more it will have to descend for knock overing; hence, more yarn if fed [3]. Tuck stitch cam provides less height as compared to knit stitch cam; thus, yarn consumption is less than knit stitch. Tuck stitches cause porosity in fabrics leading to an increase in air permeability and some other comfort properties variations. The GSM (grams per square meters/areal density) of fabric starts increasing via tuck stitches because of more yarn accumulation in the specific area. Such yarn accumulation is caused by clearing of two or more loops at a time by tuck stitches, so the point that normally consists of one yarn at a time accrues more than one yarn leading to enhanced areal density. Due to this double knocking over two yarns, loops pass from the same place in the fabric that ultimately increases the thickness of fabric in that area.

3.1.3 Miss CAM

Miss cam plate provides a straight path without any upward or downward movement; hence, the needle passes through the respective feeder without taking any height. Such straight movement does not allow the knitting needle to feed any yarn in its hook as shown in Fig. 5. Needles are responsible for constructing wales of a weft-knitted fabric. Miss stitch excludes that specific wale form the area leading to width reduction of fabric. Miss stitch consumes the least yarn among basic weft knitting stitches as no new yarn is fed at the point. Float stitches make basic fabric thinner and lighter as there are fewer loops in the fabric. Jacquard-knitted structures mostly require knit and miss stitch combination for creating visual aesthetics on fabric surfaces [4]. Similarly, fleece, knitted denim, and vertical stripes structures require miss stitch combination with knit and tuck stitches. Therefore, miss stitches are still considered important because most of the major structures are incomplete without miss stitches.

Weft knitting patterning involves needle movement on cam plate-specified tracks. Two mechanisms could be involved during knitting, one is keeping cam plates stationary and moving respective knitting needles through the cam paths. Such technique is usually followed in circular weft knitting machines with single jersey, rib, or interlock. Here, multiple feeders are involved, and each feeder is responsible for knitting a course. Needles are responsible for the number of wales knitted as the number of needles in knitting area is equal to knitted wales in specimen. The second knitting mechanism keeps the needles stationery, and cam plate moves over all needles uplifting them one by one for stitch formation. Flatbed weft knitting machines have this mechanism for stitch formation and patterning. Cam carriage setup makes oscillations over stationary needle bed for a course knitting. One oscillation knits one course in the mechanism. Productivity could be enhanced using more than one feeder in single oscillation as the number of utilized feeders is equal to the number of courses. Conventional knitting machines are of circular architecture. Stationary cam plate knitting technique is followed in the machines. Needles placed in cylinder tricks are of different butt positions, hence different cam plate track positions are provided. The average circular weft knitting machine consists of four cam tracks. At a time, one feeder works with whole needle repeat of a structural knitted cell of a course and knits that specific course on a designed repeat. Similarly, all needles pass through the next feeder knitting another course, and the cycle continues on all feeders as shown in Fig. 6.

3.2 *Jacquard Knitting*

Jacquard mechanism involves single needle control of knitting machine. Each needle from any point of knitting bed or cylinder can be selected for a specific stitch. Advanced design elements are utilized in jacquard weft knitting instead of conventionally used cam tracks and needle butt mechanism. Jacquard knitting evolved from mechanical design elements where needle disk mechanisms and pattern wheel–like design elements were used in a synchronized way to create the required pattern of fabric. Consider a pattern wheel for understanding; it consisted of ups and downs along its circumference. Knitting needles butts were contacted with those ups and downs creating cam-like pathway, needles uplifted on design height created pattern. Mechanical design elements faced certain limitations, e.g., complex design feeding system, high changeover times, and limited pattern area selection. Such limitations automatically created space for electronic elements to be used for pattern development in weft knitting jacquard machines as shown in Fig. 7.

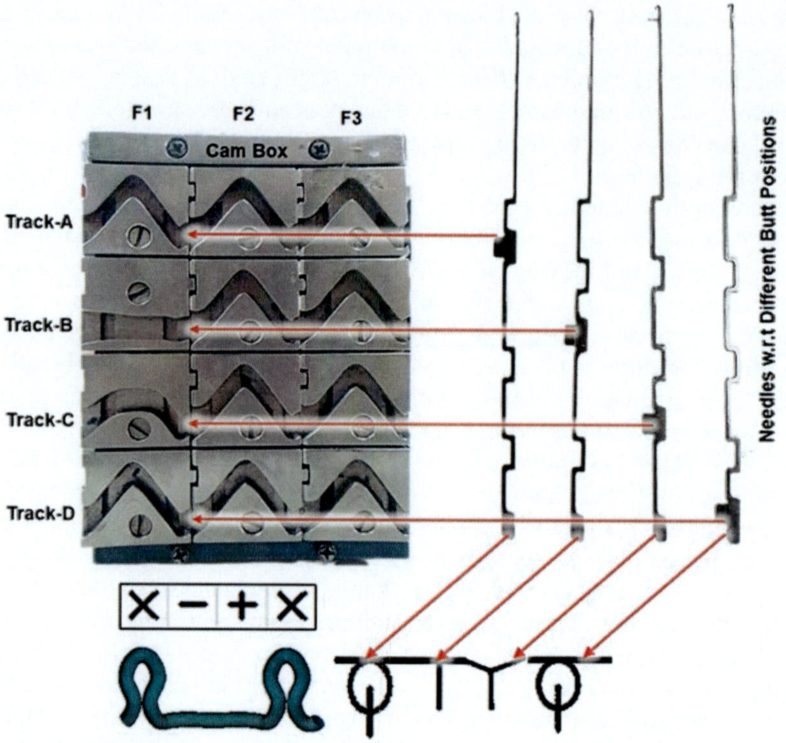

Fig. 6 CAM plate patterning mechanism

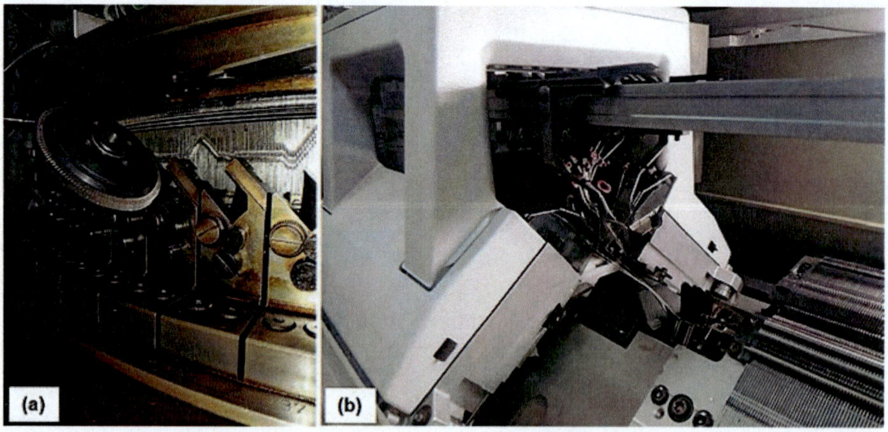

Fig. 7 Jacquard knitting elements (a) mechanical (b) electronic

Electromagnetic systems are installed nowadays in advanced jacquard machines. Knitting needles are supported by auxiliary design elements such as actuators, jacks, and selectors. One of the auxiliary elements receives electromagnetic signal, and in response, vertically corresponding needles are uplifted. In jacquard flatbed, weft knitting cam carriage system provides movement to needles on the required jacquard pattern, as fundamental jacquard elements are installed inside the carriage system [5]. Electronically controlled jacquard weft knitting machines have design development software, e.g., APEX. Knitting patterns are developed on software and loaded on machines via USB devices.

4 Single Jersey Derivatives

4.1 Plain Jersey

Plain jersey fabric consists of all knit loops in its construction. Knit loop consumes most yarn during construction as compared to other stitches, hence loop architecture makes plain jersey fabrics stretchable among other single derivatives as shown in Fig. 8. Plain jersey fabric requires least machine modifications/alterations for its formation. The structure is mostly preferred for innerwear due to its good stretch properties making it comfortable to wear. Woven fabrics have only a crimp factor governing stretch, while weft-knitted fabric loop structure distributes yarns in both axial and lateral directions. Thus, stretch properties of weft-knitted fabrics are higher up to 100% rather than 10% of woven fabrics. Yarn consumption factor elaborates fabric stretch. Plain jersey structure consisting of all knit stitches has superior stretch due to more available yarns, and consequently recovery properties are also viable. Axial stretch and recovery are lower widthwise due to more yarn contributions in width. GSM indicating grams in a square meter of fabric is influenced by knitting structural parameters. Shrinkage of weft-knitted fabrics is also structure dependent. Plain jersey fabrics have higher widthwise shrinkage caused by knit loops. Moreover, plain jersey has a more snarling phenomenon. Tuck stitches cause yarns to accumulate in a specific area, hence GSM is increased via tuck stitches. Plain jersey exhibits normal GSM values due to all knit stitches. Fabric thickness

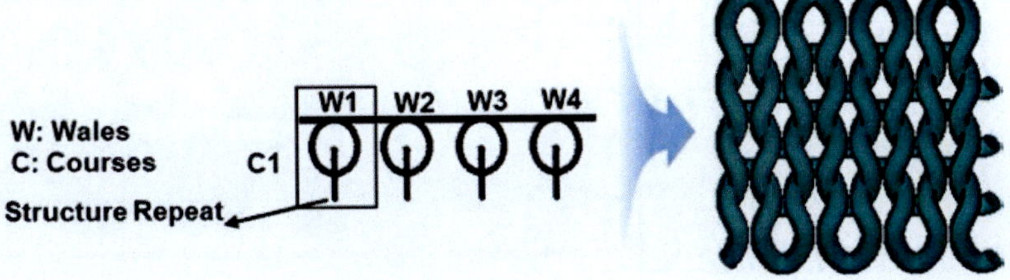

Fig. 8 Plain jersey

and compression properties are directly related. Plain jersey fabrics have the least thickness. However, the thickness is still three times the diameter of yarn. Bursting strength is affected by yarn orientations and consumption in weft-knitted structures. Plain jersey fabrics have the highest bursting strength among single jersey weft-knitted fabrics. Air permeability, thermal conductivity, and water vapor permeability of plain jersey fabrics are of optimum range, i.e., in between structures consisting of tuck and miss stitches. Serviceability, pilling, and abrasion properties of plain jersey fabrics are poor due to more available surface and loose structures. The properties could be enhanced using tighter knits and high-twist yarns, but tactile comfort properties are then compromised.

4.2 Single Locast Pique

Single Locast pique structure often termed PQ fabric is a combination of knit and tuck stitches. Verbally saying, it has first course of all knit stitches, second course of knit on odd and tuck on even needles, third course of all knit stitches, and fourth course of tuck on odd and knit on even needles. The structure has design repeat of four courses and two wales as shown in Fig. 9. Hence, using conventional circular weft knitting machine, it will acquire minimum of four active feeders for its production. Tuck stitches cause porosity in the fabric that opens pathways for air permeation. This is why PQ fabrics have excellent air permeability and are mostly preferred in summer wear T-shirts. Stretch and recovery properties of single PQ structure are lesser than plain jersey fabrics due to tuck stitches consuming less yarns. Weft-knitted fabrics having tuck stitches are known to offer higher lengthwise shrinkage. Tuck stitches are responsible for accumulating yarns in specific areas of fabric, which in turn cause increase in GSM, and air permeability, water vapor permeability, and thermal conductivity are enhanced due to induced porosity [6]. Thickness and compression of single PQ structure is more than plain jersey. While bursting strength is lower than plain jersey fabrics and miss stitches

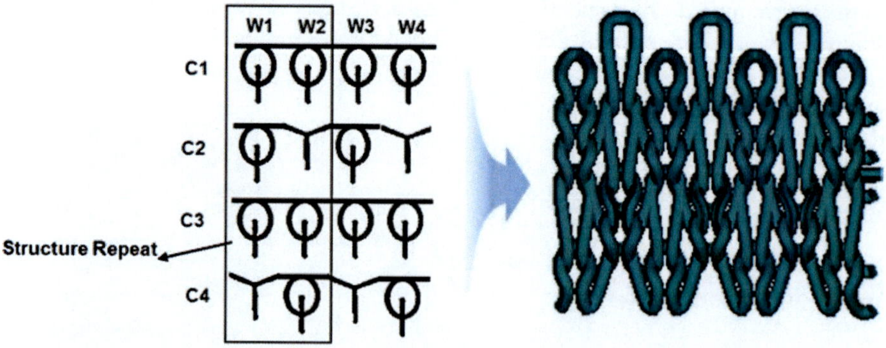

Fig. 9 Single locast PQ

comprising structures due to less yarn consumption, and limited yarn orientation geometries [4]. Tactile comfort of single PQ fabric is less than plain jersey and miss stitch structures, i.e., cross miss, and fleece.

4.3 Double Locast Pique/Double Cross Tuck 1 × 1 and Jumbo PQ

The structure is further modification/expansion of single PQ fabric. On expansion, structural repeat goes on six courses, but the number of wales in a repeat remains the same as single PQ. Due to the addition of more tuck stitches than single PQ as shown in Fig. 10, the structure exhibits good air permeability providing thermal comfort. Hence, the structure is also preferred while fabricating summer clothing. Air permeability, water vapor permeability, thermal conductivity, and all other fluid conduction properties of double cross tuck are greater than single PQ and plain jersey fabrics. More tuck stitches are responsible for more lengthwise shrinkage in double cross-miss structure. GSM has enhanced values than single PQ and plain jersey structures. However, bursting strength and tactile comfort are compromised, and tuck stitches lead to limited yarn geometries. Increasing stitch density can enhance bursting strength, but comfort properties seem to be slashed.

The jumbo pique structure is wale-wise extension of double PQ structure. Wale repeat of jumbo pique structure moves onto four needles/wales than two in a single pique. Courses having all knit stitches remain the same. However, courses having knit and tuck stitches show alteration of stitches in course after two needles instead of one, e.g., two knit and two tuck as shown in Fig. 11. The addition of tuck stitches

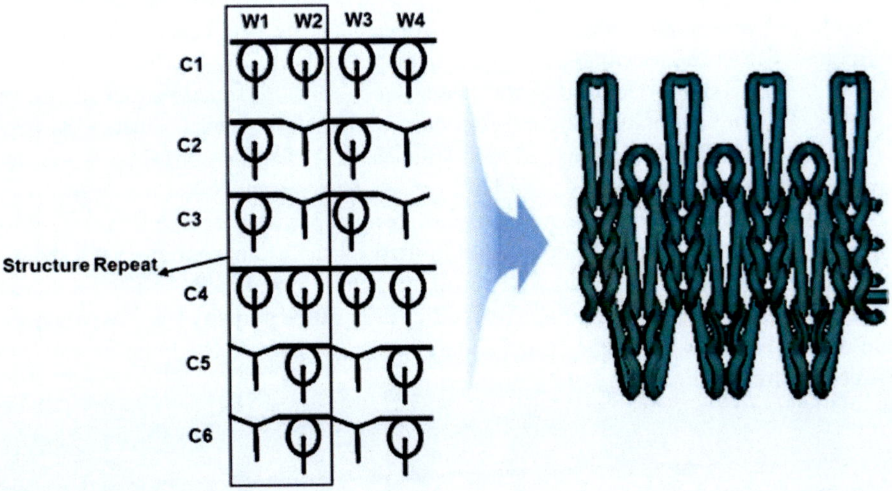

Fig. 10 Double locast PQ

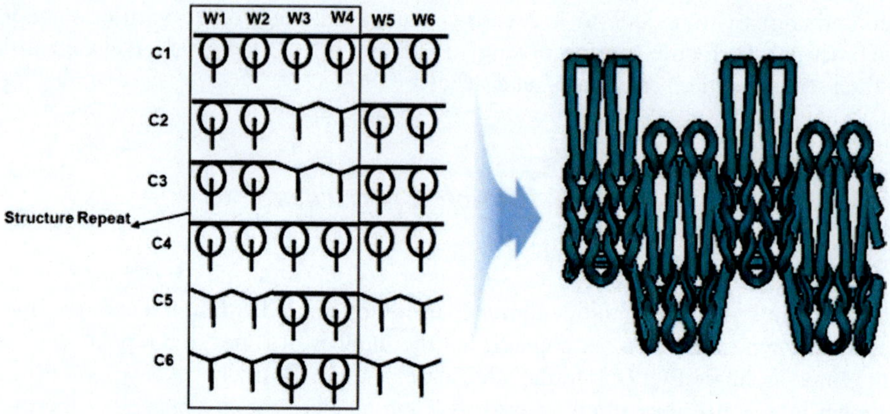

Fig. 11 Jumbo PQ

further increases the properties following an increasing trend, and in a similar way, properties following decreasing trend are further decreased. End uses of jumbo pique are the same as single and double pique.

4.4 Honeycomb and Jumbo Honeycomb

Honeycomb structure is also a combination of knit and tuck stitches, but the structural repeat is of smaller unit as compared to pique structures. Verbally, it has a structure repeat of two wales and two courses. The first course consists of knit stitches on odd needles and tuck stitches on even needles. The second course comprises tuck stitches on odd needles and knit stitches on even needles as shown in Fig. 12a. Jumbo honeycomb structure is formed by the extension of honeycomb structure. Structural repeat goes on four courses and four wales. Verbally, alteration of stitches in a course occurs after two needles, e.g., two knit, and two tuck as shown in Fig. 12b. And alternation of courses occurs after each two courses instead of one. Honeycomb derivatives consist of tuck stitches at nearer positions than PQ structures, as all knit course is absent in honeycomb. Honeycomb structures seem to be rigid and harsh, hence stretch and tactile comfort properties are not much extraordinary. However, recovery of honeycomb-knitted fabrics is appropriate. Lengthwise shrinkage is severe, while widthwise shrinkage is minimized. GSM is also increased due to more yarn's accumulation in a specified area. Huge porosity leads to enhanced air permeability, water permeability, and thermal conductivity. But bursting strength is compromised.

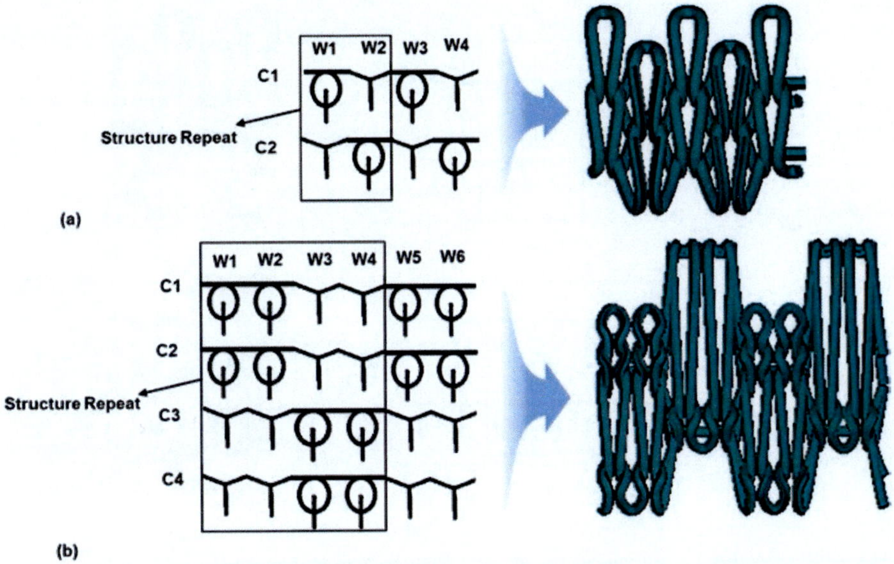

Fig. 12 Honeycomb derivatives (**a**) honeycomb (**b**) jumbo honeycomb

4.5 Single and Double Cross Miss

Single cross-miss structure is a combination of knit and miss stitches. Just like honeycomb structure, its repeat goes on two courses and two wales. However, the tuck stitches are replaced by miss stitches as shown in Fig. 13a. Verbal notation demonstrates it as a structure consisting of the first course with knit stitches on odd needles and miss stitches on even needles while the second course with miss stitches on odd needles and knit stitches on even needles. Double cross miss is also a combination of knit and miss stitches. However, structural repeat of the design goes on needles instead of two single cross miss keeping course repeat of two as shown in Fig. 13b. Verbally, the structure has two course repeats. The first course consists of knit stitches on first two needles (one and two) followed by two miss stitches on the upcoming two needles (three and four). The second course consists of miss stitches on the first two needles (one and two) and knit stitches on next two needles (three and four). Miss stitch consumes the least yarn; hence stretch is lower. However, recovery is viable. Cross-miss structures are thinner in width than plain and tuck stitch comprising structures, leading to widthwise shrinkages. GSM of cross-miss structure lies in between plain jersey, PQ and honeycomb structures. Miss stitches appear as floats on fabric's technical back, governing an increase in GSM. Air permeability, water vapor permeability, and thermal conductivity have medium/optimum values. Tactile comfort of cross-miss structural derivatives is considered superior to plain jersey because of surface smoothness, bulkiness, and better compression properties. Historically, bursting strengths of cross-miss structures are higher than PQ and honeycomb structures, while inferior to plain jersey.

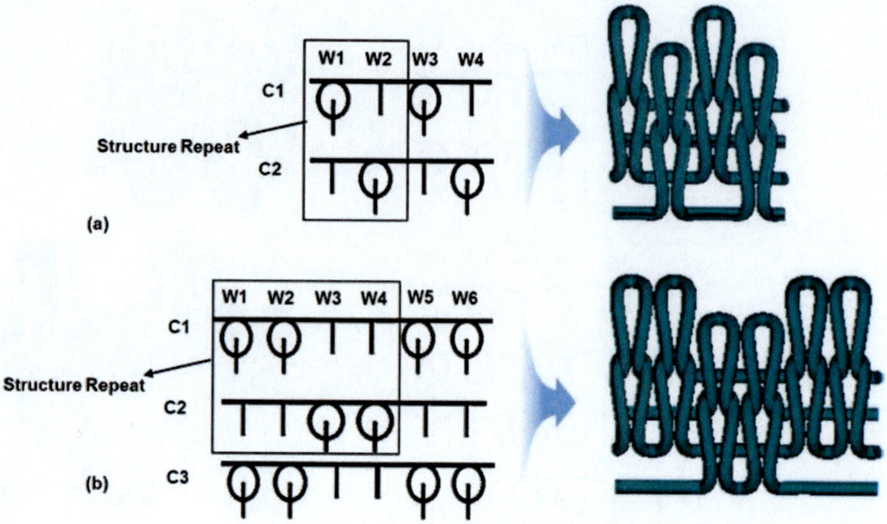

Fig. 13 Cross-miss derivatives (**a**) single cross miss (**b**) double cross miss

4.6 Two-Thread Fleece Derivatives

Fleece fabrics are conventionally called fabrics having raised surfaces on at least one fabric side. However, such a raised surface is achieved through an additional raising process after knitting. Fleece structures are knitted in such a way that floating yarns appear on the backside of the fabric. Those floats are then sheared resulting in a fluffy appearance. Focusing on knitting structure, two-thread fleece fabric has a course repeat of two needles, and normally wale repeat is of four needles. The first course of two-thread fleece is always of all knit stitches. However, the second course consists of multiple combinations of tuck and miss stitches (Fig. 14). Fleece fabrics are preferable for winterwear, i.e., hoodies and sweatshirts. Fleece fabrics exhibit viable thickness as compared to all other single jersey derivatives. Shearing process adds a positive impact on thickness and compression properties. Bursting strength becomes lower, raising process damage fibers. Stretch and recovery properties of fleece derivatives are low to medium due to the fabric's architectural influence. GSM of fleece fabric is the highest among all single derivatives. Tuck/loop yarns of fleece are of courser counts making GSM values higher. Air permeability, water vapor permeability, and thermal conductivity are lowest among single jersey derivatives. However, thermal insulation/resistance properties are noticeable, making all fleece derivative suitable for warm thermo-physiological and tactile property achievement. Shrinkage is an adverse factor of two-thread fleece. Higher shrinkages after relaxation lead to serviceability issues. Hence, three-thread fleece derivatives are engineered, which are termed low shrinkage fleece (LSF).

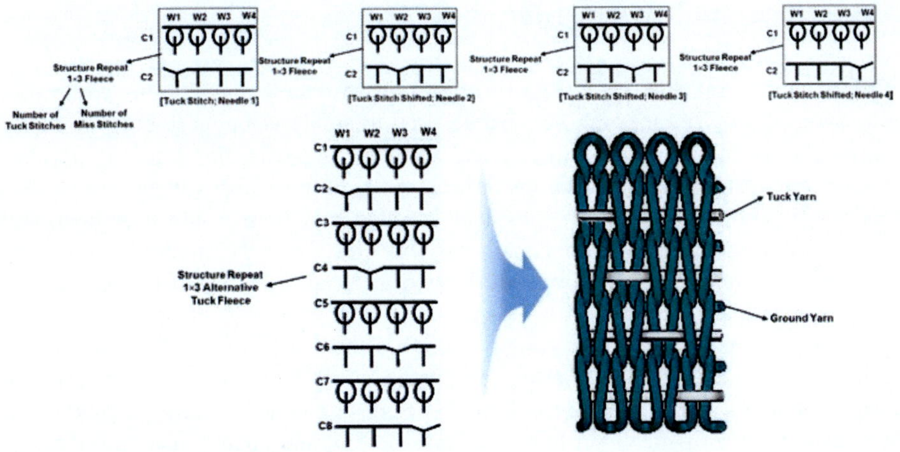

Fig. 14 Two-thread fleece structures

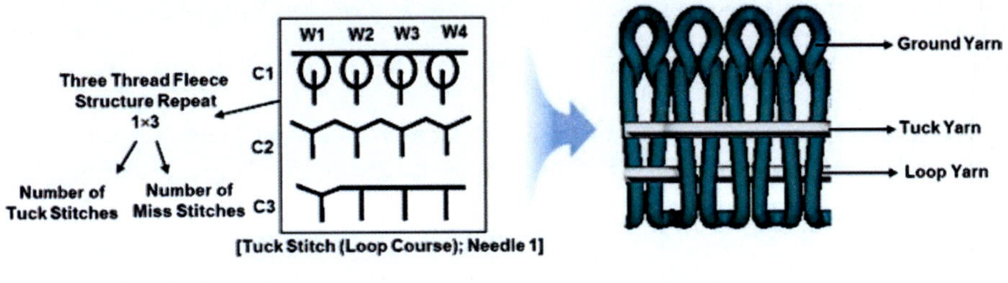

Fig. 15 Three-thread fleece

4.7 Three-Thread Fleece Derivatives

Three-thread fleece comprises an additional all-tuck course with previously existing knit and loop courses. The first course is of all knit stitches, second course of all tuck stitches, and third course of tuck and miss stitches. Tuck and miss stitch course are also termed loop or draft course and is shown in Fig. 15. Three-thread fleece is conventionally termed low shrinkage fleece (LSF). Usually, polyester yarn is used in tuck courses, and during heat setting, polyester becomes set on required dimensions governing less shrinkage. Physical, mechanical, thermos-physiological, and tactile comfort properties are the same as two-thread fleece with an additional aspect of low shrinkage.

4.8 Horizontal Stripes Fabrics

Stripes fabrics fall into the category of knitted structures and designs with color variations. Novel color combinations are used to create visual aesthetics on fabric surface as shown in Fig. 16. Such visual aesthetics majorly play a role in psychological comfort providing soothing and relaxed feeling to both viewer and wearer. Except psychological comfort, physical, mechanical, thermos-physiological, and tactile comfort properties are like plain jersey fabrics. From the manufacturing point of view, horizontal stripes fabrics are knitted using all knit cams on circular weft knitting machines. Number of feeders are allocated to specific colors, which in turn produce required colored courses governing horizontal stripes in the weft-knitted fabric. On conventional weft knitting machines, colors can only be adjusted till the total number of feeders on the machine, hence design repeat remains limited. Auto striper weft knitting machines had overcome the limitation using feeders and finger mechanisms. Each feeder further comprises specific number of fingers consisting different color yarns. Electromagnetic system controls yarn feeding creating more opportunities of design enhancement.

4.9 Vertical Stripes Fabric

Weft-knitted vertical stripes fabrics are designed using different derivatives of cross-miss designs. Properties of vertical striped fabrics are identical to cross-miss single jersey derivatives. Knit and miss stitch combinations are alternatively used in

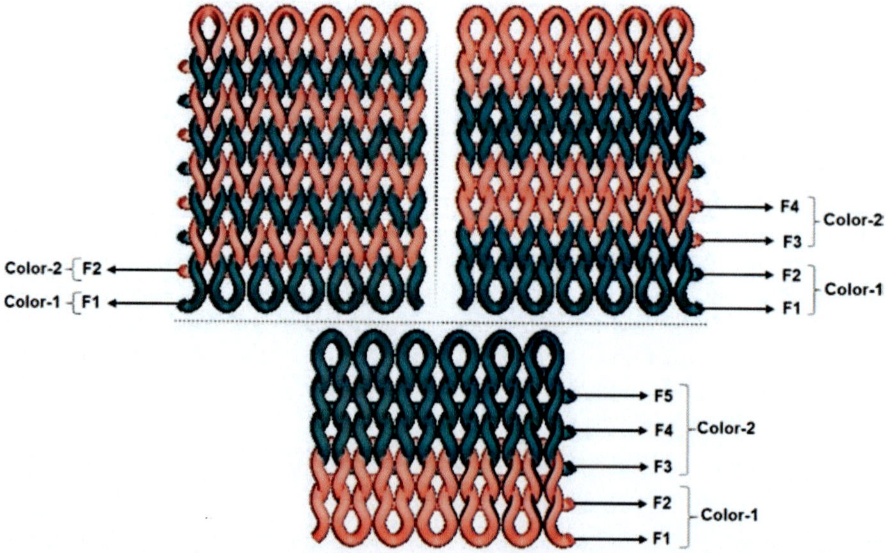

Fig. 16 Horizontal stripes

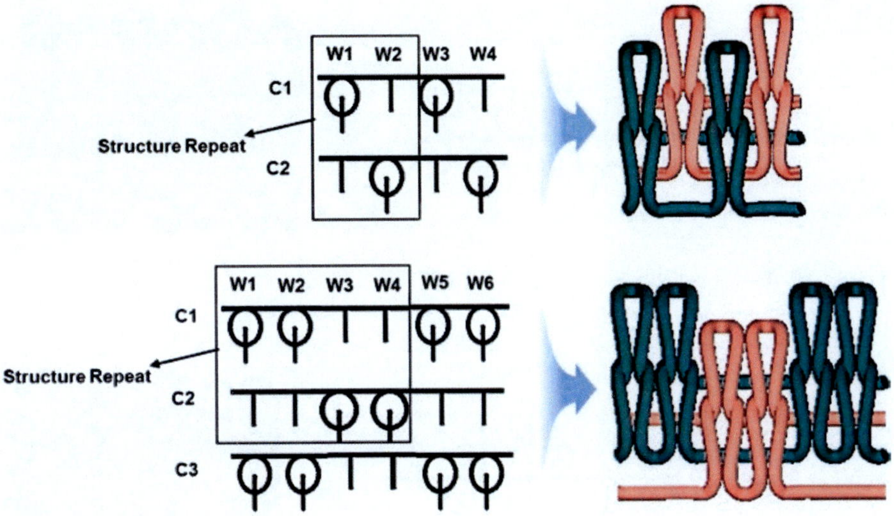

Fig. 17 Vertical stripes

courses and wales to create different colored stripes as shown in Fig. 17. Vertical stripes are responsible of creating surface visual aesthetic properties on fabric surface in the vertical direction.

5 Rib Structural Derivatives

Rib structures are engineered on weft knitting machines having dial and cylinder with alternative needle gating. Alternative needle gating allows lifting of both dial and cylinder needles simultaneously. Hence, one complete revolution governs one course. Further characteristics of rib structural derivatives are described in this section.

5.1 Plain Rib

Plain rib fabric comprises alternative knit stitches of dial and cylinder. In flatbed knitting machines, the knitting needles of dial and cylinder are replaced by back and front bed needles for rib structure formation as shown in Fig. 18. Inherently, plain rib structure processes the highest stretch and recovery properties among all weft-knitted structures. Hence, rib structures are mostly used in collars, arm welts, and waist welts of hoodies and jackets. GSM of rib derivatives lies in between plain jersey and interlock derivatives, higher than plain jersey, and lower than interlock fabrics. Thermal insulation of rib fabrics are higher among weft-knitted architectures. Tactile comfort properties are of average range. Thickness and compression

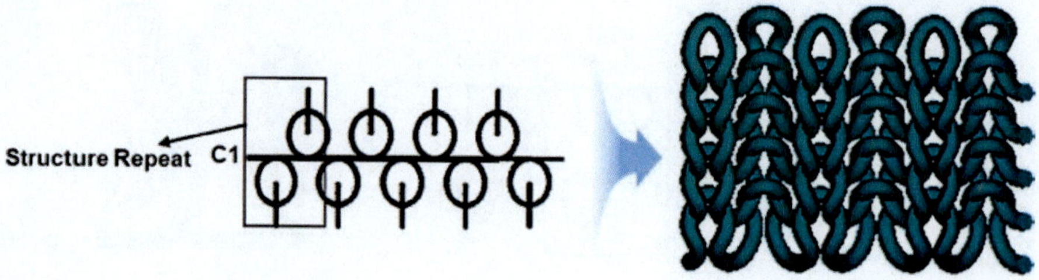

Fig. 18 Plain rib 1 × 1 structure

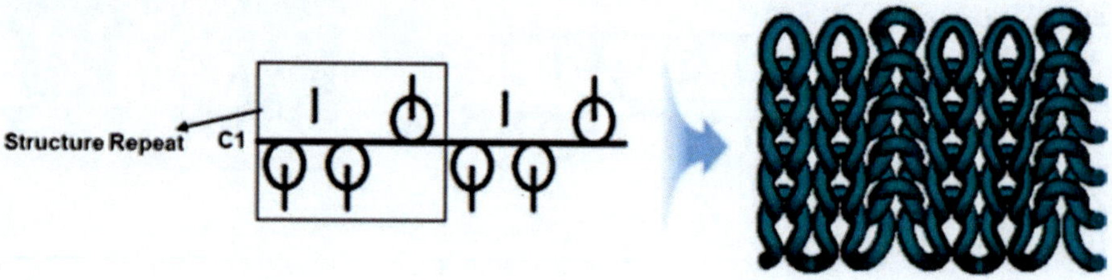

Fig. 19 Rib 2 × 1 structure

properties are higher. Shrinkage being the inherent property of weft-knitted fabrics exists in plain rib. However, snarling (curling of fabrics from edges) is exempted due to balanced loops on both fabric faces. Bursting strength is noticeable, and higher strain values are governed by better stretch.

5.1.1 Plain Rib Derivatives

Plain rib structures can be modified using different combinations of dial and cylinder needles, i.e., one dial needle knits after each two knit stitches of cylinder (2 × 1 rib) as shown in Figure 19. Similarly, numerous combinations can be utilized governing unique fabric architectures and properties related to thermo-physiological and psychological comfort. Historically, increasing rib order from 1 × 1 to greater i.e., 2 × 2 increases air permeability and moisture transportation properties, while 1 × 1 rib structures is best for thermal insulation, hence used in winterwear.

5.2 Milano Rib Derivatives

Milano structural derivatives are specific combinations of knit and miss stitches of dial and cylinder. In alternative half-Milano rib structure, the first course is of all knit stitches on dial and cylinder; the second course has knit stitches on cylinder

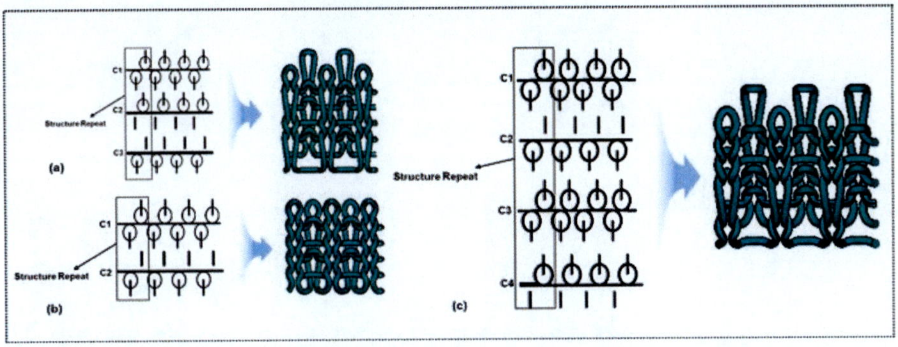

Fig. 20 Milano derivatives (**a**) alternative half Milano (**b**) half Milano (**c**) Milano rib

with all needles of cylinder miss; the third course is of plain knit on dial and cylinder; and the fourth course is of all dial knit and cylinder miss. Half-Milano rib pattern is a combination of knit and miss stitches. It has a structural repeat of two courses. The first course has all knit on both dial and cylinder, while the second course has all knit on cylinder with all miss stitches on dial. Milano rib is a further extension of half-Milano rib. Course repeat of structure moves on three feeders. The first course is of all knit stitches on dial and cylinder; the second course is of all knit stitches on dial with all cylinder needles miss; the third course is of all knit stitches on cylinder; however, dial has all miss stitches (Fig. 20). Milano rib derivatives have lower GSM values than plain rib fabrics. Miss stitches govern width reduction phenomenon, leading to more widthwise shrinkage in Milano derivatives. Stretch and recovery properties of Milano rib derivatives are less than plain rib structure as miss stitches engineered by less yarn consumption cannot contribute in an appropriate way for stretch and recovery of fabrics. Air permeability, water vapor permeability, and thermal conductivity are enhanced due to thinning of fabric. Tactile comfort properties of Milano rib are superior to plain rib. However, bursting strength of Milano rib derivatives is inferior to plain rib owing to miss stitches.

5.3 *Cardigan Rib Derivatives*

Cardigan rib structure is a combination of alternative knit and tuck stitches on cylinder and dial. In the first course, the cylinder has all knits, and dial has all tuck stitches (Fig. 21). While in the second course, the cylinder has all tuck stitches and dial knits. Tuck stitches in the structure increase GSM and porosity of structure. The pores allow more air transportation enhancing thermo-physiological comfort in warm weather. Lengthwise shrinkage occurs by tuck stitch induction. Bursting strength is experienced lower for cardigan rib structure as tuck stitch structures have less mechanical strength than knit and miss stitches comprising structures. Double cardigan structure is a further modification of cardigan structure. The structure

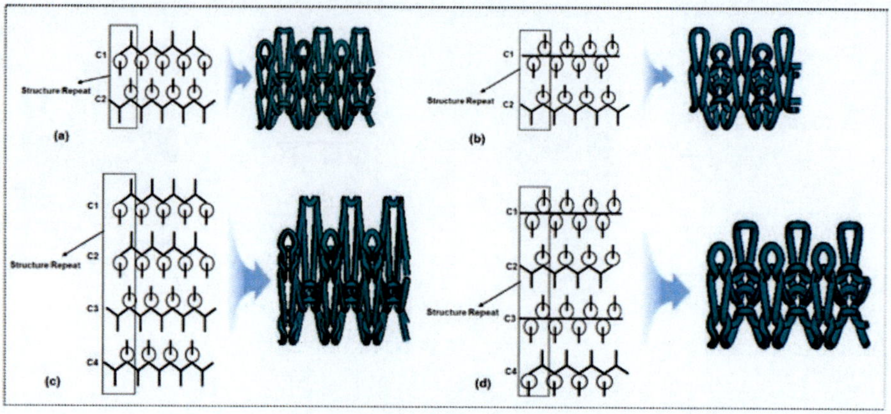

Fig. 21 Cardigan derivatives (**a**) cardigan (**b**) half Cardigan (**c**) double ccardigan (**d**) half-cardigan double sided

consists of design repeat of four courses and two wales. The first two courses are of cylinder knit and dial tuck. While the second two courses are of dial knit and cylinder tuck. Consecutive vertical tuck stitches are introduced in the structure for both dial and cylinder alternatively. Hence, the pore size due to tuck stitches is increased governing more air permeability than cardigan. GSM becomes higher than simple cardigan structures. However, tuck stitch–dependent properties increase or decrease in their linearly followed trend. Half cardigan consists of plain one-by-one rib course and cardigan design combination. The first course is of all dial and cylinder knit stitches. And the second course is of all knit on dial and all tuck stitches on cylinder. In such pattern, tuck stitches appear on the face side of the fabric allowing more air to permeate from inside to outside. Porosity of half-cardigan structure is less as compared to cardigan structure, hence thermo-physiological comfort properties feasible for warm weather are inferior. Double-side half-cardigan structure is a modified form of half-cardigan structure. Wale repeat is of two needles like half cardigan, while course repeat goes on four courses. First and third courses are of all knit stitches. In the second course, cylinder has all tuck, and dial has all knit stitches. However, in the fourth course, cylinder has all knit, while dial has all tuck stitches. In this manner, the number of tuck stitches becomes balanced and more than half cardigan. Hence, summer weather thermo-physiological comfort properties are more than half cardigan and less than simple cardigan structure.

5.4 *Rib Ripple*

Rib ripple is a combination of knit and miss stitches of a specific pattern on cylinder and dial. Course repeat comprises three feeders. The very first course is of all knit stitches on dial and cylinder. However, the second and third courses have all knit on cylinders and miss stitches on dial (Fig. 22). Miss stitch induction governs thin

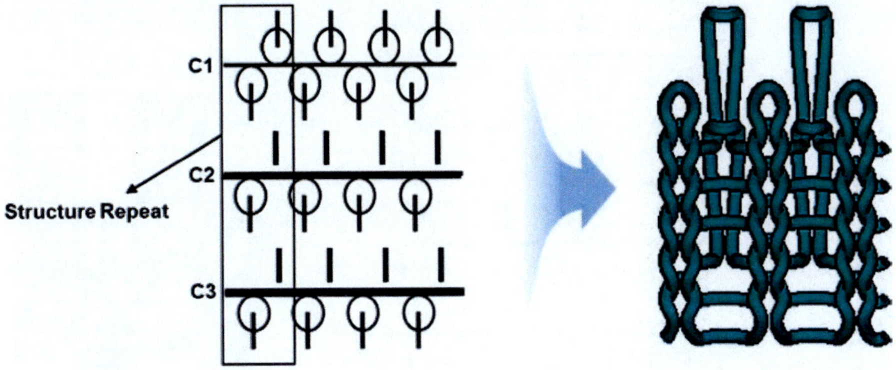

Fig. 22 Rib ripple

fabrics, and warm weather thermo-physiological comfort properties are enhanced/ greater than plain rib but remain lower than cardigan rib derivatives. Widthwise shrinkage of rib ripple is higher via miss stitches. However, bursting strength of rib ripple structure is higher than cardigan rib derivatives. Similarly, air permeability, water vapor permeability, thermal conductivity, and thermal insulation values are comparable to Milano rib structural derivatives.

5.5 *Belgian Double PQ*

Belgian double PQ is a rib-engineered fabric with a combination of knit and miss stitches uniquely on dial and cylinder. Course repeat is of six feeders, while wale repeat comprises two needles of each dial and cylinder (Fig. 23). Like single jersey PQ, the structure possesses pores allowing more fluid transportation, which leads to enhanced air permeability, water vapor permeability, and thermal conductivity properties. Tactile comfort is also experienced optimum. GSM lies above plain rib structural derivatives. Due to miss stitches, widthwise shrinkage dominates over lengthwise. However, the bursting strength of Belgian double PQ is higher than that of conventionally knitted PQ structures, as tuck stitches are replaced by knit stitches in the structure.

6 Interlock Structural Derivatives

Interlock weft-knitted fabrics are formed on double-needle bed machines. Rib and interlock weft knitting machines differ in needle gating. For interlock fabrics, needles are exactly upon each other. To avoid crashing of needles exactly upon each other during knitting, needles are alternatively raised to facilitate fabrication. Interlock derivatives are also formed via different unique combinations of knit,

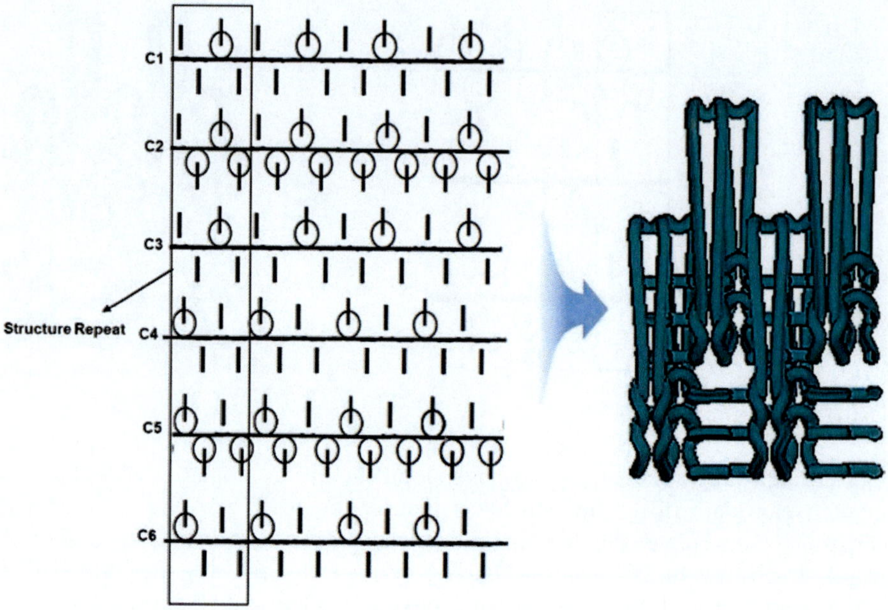

Fig. 23 Rib Belgian double PQ

tuck, and miss stitches. Such combinations are responsible for inducing required thermos-physiological, physical, and psychological comfort properties in fabrics.

6.1 Plain Interlock

Plain interlock structure is engineered by alternative knitting of dial and cylinder needles. Two feeders are responsible for one course knitting in interlock structures. The first feeder knits alternative half needles of dial and cylinder, and the second feeder is responsible for the remaining half needles knitting (Fig. 24). Thickness and compression properties of interlock knits are superior to all weft knitting structural derivatives. Surface smoothness enhances tactile comfort properties. Interlocking of dial and cylinder loops with each other governs structure having viable bursting strength. Air pockets between both knitted layers provide a modified thermal insulation, hence interlock fabrics are suitable for warm clothing.

6.2 Interlock Cross Tubular

Interlock cross tubular fabrics consist of vertical tube structure. Instead of one-by-one needle alteration, there are two-by-two or three-by-three needle alternative knitting forming tubes in wale-wise fabric direction (Fig. 25). Such tubes are useful for

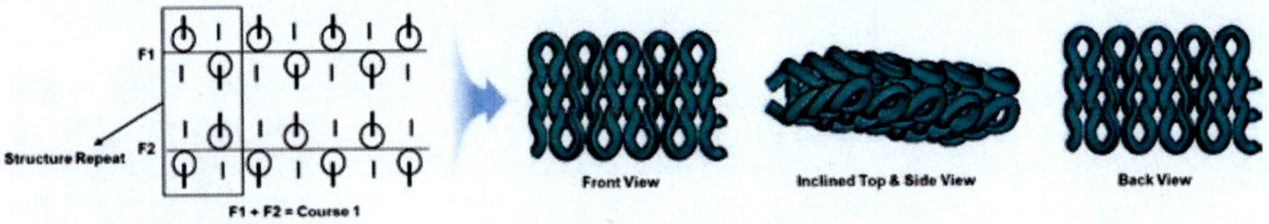

Fig. 24 Plain interlock

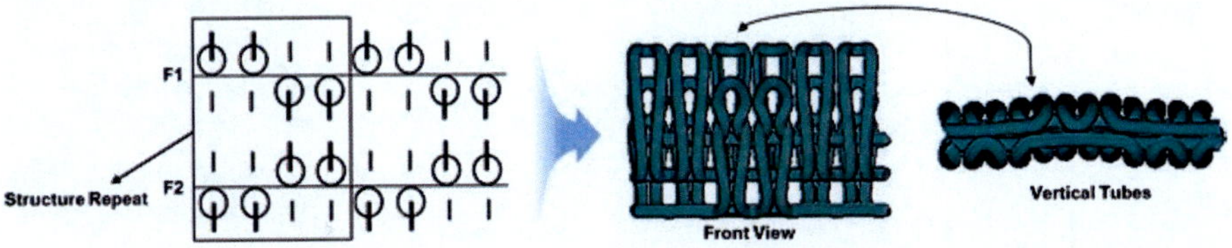

Fig. 25 Interlock cross tubular

integrating circuits and other auxiliary elements necessary for electronic integration in smart knitted garments. However, due to increased needle alteration number, thermal insulation is somehow compromised, while air permeability and water vapor permeability are enhanced via reduced structure compactness. Tactile comfort is inferior for interlock cross tubular; more dumps on fabric reduce surface smoothness.

6.3 Interlock Half Cardigan

Interlock half cardigan's stitch architecture is the same as rib half cardigan. However, the final fabric aspects are quite different due to needle gating differences. Feeder repeat of the structure goes on four feeders, and the number of knitted courses is two for one repeat (Fig. 26). Tuck stitches are responsible for porosity, and yarn accumulation in specific areas of fabric tends GSM to go on higher values. However, enhanced porosity increases air permeability, water vapor permeability, and thermal conductivity of interlock half cardigan. Higher lengthwise shrinkage is experienced, as tuck stitches in the structure show shrinkage along wale direction. Tactile comfort is less due to the crispy surface. Thickness and compression lie in the medium to higher range for interlock structures. Bursting strength of interlock is lower than miss stitches comprising interlock structures, i.e., interlock Milano and interlock modified.

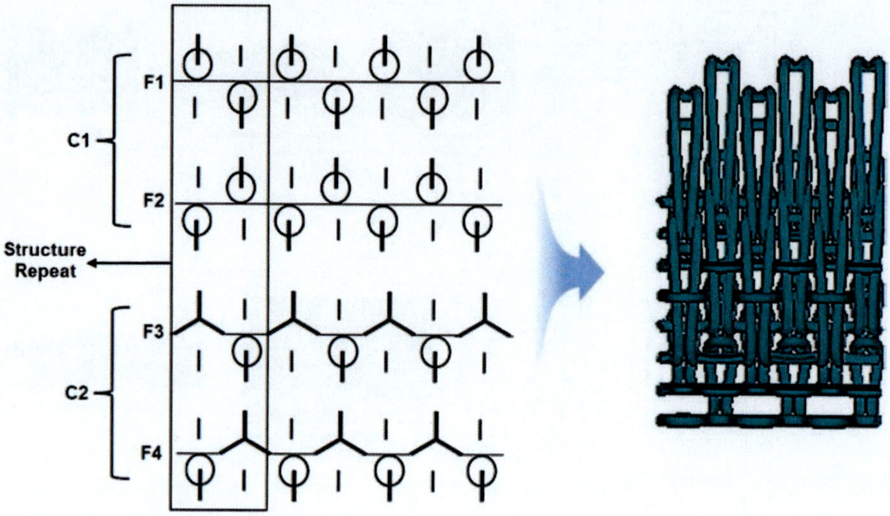

Fig. 26 Interlock half cardigan

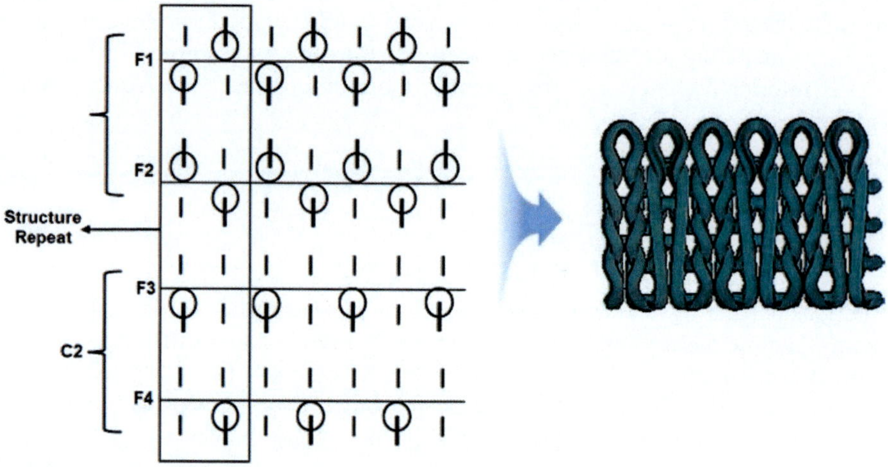

Fig. 27 Interlock half Milano

6.4 *Interlock Half Milano*

Similar to rib Milano structural derivatives interlock half Milano is the combination of knit and miss stitches in a unique manner on dial and cylinder. Structure knitting repeat consists of four feeders and consequently of two courses (Fig. 27). Miss stitches tend fabric to exhibit widthwise shrinkage. Thickness and compression properties are lower than other interlock derivatives. Thin fabric architecture leads to good air permeability, water vapor permeability, and thermal conductivity. Tactile comfort of interlock Milano lies between plain interlock and cardigan structures, higher than cardigan and lower than plain interlock. Bursting strength of interlock Milano is also higher than cardigan derivatives due to miss stitches, while inferior to plain interlock structure.

6.5 *Interlock Modified*

Interlock modified is a unique combination of plain interlock course and all knit cylinder course. Knitting structural repeat goes on four feeders (two courses). Horizontal thin lines are experienced in whole fabric via one course consisting of knit on cylinder only and all miss stitches on dial (Fig. 28). GSM of modified interlock is reduced, and miss stitches consume less yarn. Fabric shrinks widthwise. Tactile comfort also lies in optimum range. However, air permeability, water vapor permeability, and thermal conductivity are experienced better due to thinning of fabric. Thickness and compression properties are inferior to other interlock structural derivatives; miss stitches reduce thickness, and reduced thickness governs poor compression characteristics.

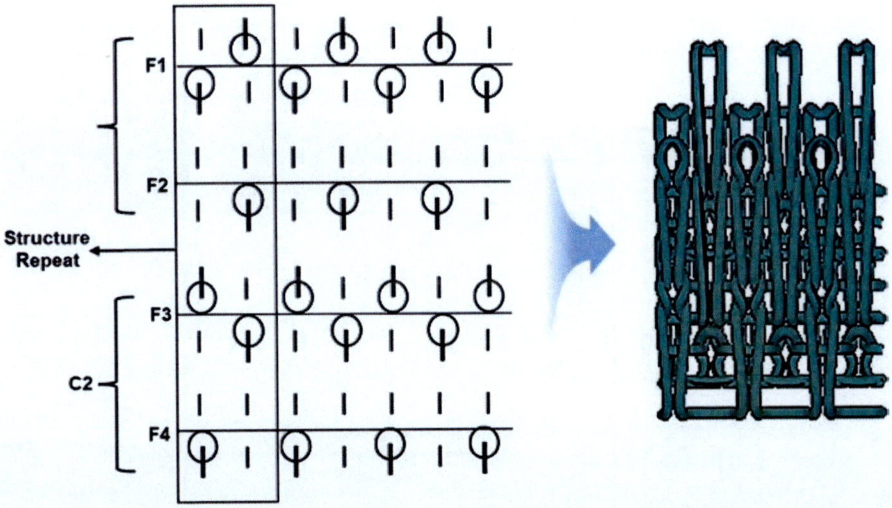

Fig. 28 Interlock modified

7 Specialty Knitted Structures

7.1 Plaited Fabrics

Plaited weft-knitted structures are plain single jersey fabrics having elastane yarn knitted in each course. Elastane yarns are composed of polyurethane segments possessing higher stretch and recovery characteristics. Hence, plaited structures are preferred in high-stretch applications, where body fitting is a primary factor, i.e., sportswear. Structurally saying, knitting repeat of plaited structures is of a knit loop with an additionally fed naked elastane along with main yarn (Fig. 29). Tactile comfort is comparable to plain jersey fabrics. Bursting strains of plaited knitted structures are higher due to elastane. However, thermos-physiological comfort is compromised via poor fluid transportation characteristics of plaiting yarns.

7.2 Three-Dimensional Spacer Fabrics

Three-dimensional weft-knitted fabrics consist of yarns in x, y, and z axes. Thickness/height of fabrics is considerable along with length and width parameters. Such knitting is only possible with weft knitting machines having jacquard facility, i.e., single-needle selection, and more than one needle bed in most of the cases. 3D-spacer fabrics are among well-known examples of three-dimensional weft-knitted structures. Spacer fabric consists of three layers (Fig. 30), top and bottom base knitting layer and a central joining layer. Top and bottom layer could be of any single jersey-engineered architecture. However, the joining layer is a monofilament

Fig. 29 Plaiting structure

Fig. 30 3D spacer fabric

having alternative tuck stitches on both needle beds. Mechanical properties, i.e., bursting strength, are higher as compared to conventional two-dimensional knitted fabrics. Dimensional shrinkages are negligible as knitting yarns are balanced in 3D grid, hence they are locked at their adjacent points. Tactile properties are noticeable due to compression property contribution. Compression properties of spacer fabrics are noticeable, thus spacer fabrics have the potential to replace polyurethane foams in the future. Good air permeability, water vapor permeability, and thermal conductivity enable space fabrics to be used in cushioning applications with enhanced thermos-physiological comfort.

7.3 Intarsia Technique

The intarsia technique ensures control over single knitting loop instead of single needle control in jacquard weft knitting. Intarsia facility is available on advanced flatbed weft knitting machines. Special stoppers are mounted on carriage rails. Yarn carriers are selected only on required knitting area and are stopped after the point. New carrier is selected from the point to knit, and similar engineering continues. Such selection and stopping mechanism allow complex knitting patterns without having floats on technical backside of fabrics, as experienced in jacquard knits. Mechanical and thermos-physiological properties of intarsia knitted structures are comparable to plain jersey structures. Hence, intarsia technique allows complex designs/shape knitting without having floats issue and keeping physical properties similar to plain jersey.

References

1. D.J. Spencer, *Knitting Technology: A Comprehensive Handbook and Practical Guide*, 3rd edn. (Woodhead Publishing, 2001)
2. Y. Nawab, S.T.A. Hamdani, K. Shaker, *Structural Textile Design. Structural Textile Design* (2017)
3. C. Iyer, W.S. Bernd Mammel, *Circular Knitting*, 2nd edn. (Mesienbach, Bamberg, 2004)
4. K.F. Au, *Advances in Knitting Technology*. 1st ed. (E. Manchester, ed., Woodhead Publishers, Textile Institute, 2011). 318 p
5. S.C. Ray, *Fundamentals and Advances in Knitting Technology* (Woodhead Publishing Limited, 2011)
6. N. Emirhanova, Y. Kavusturan, Effects of knit structure on the dimensional and physical properties of winter outerwear knitted fabrics. Fibres Text. Eastern Europe **16**(2), 69–74 (2008)

Mechanics of Weft-Knitted Structure

Hafsa Jamshaid and Rajesh Mishra

1 Mechanics of Weft-Knitted Structures

1.1 Introduction

Knitting is the second most popular fabric formation technique by interloping or intermeshing of only one set of yarn. A straight yarn is provided to the needle and it converts it into a vertical set of intermeshed loops. The word "knitting" is derived from the Saxon word "Cnyttan," which in turn is derived from the ancient Sanskrit word "Nahyat." Knitting can be classified into two types according to the direction of formation of loops.

- Weft knitting.
- Warp knitting.

This chapter focuses on the mechanics of weft-knitted structures. Weft knitting is a technique in which a loop is formed horizontally and a single set of yarn is required for loop formation. Knitted fabrics are very famous for their good extensibility up to 200% depending upon the type of structure. Weft-Knitted fabrics are known for their properties such as good stretch, flexibility, comfort and soft hand feel. On the contrary, they also have some disadvantages such as low strength, shrinkage and growth problems. It is generally considered that knitted fabrics are

H. Jamshaid (✉)
School of Engineering and Technology, National Textile University, Faisalabad, Pakistan
e-mail: hafsa@ntu.edu.pk

R. Mishra
Faculty of Engineering, Czech University of Life Sciences Prague, Prague, Czech Republic

H. Jamshaid, R. Mishra (eds.), *Knitting Science, Technology, Process and Materials*,
Textile Science and Clothing Technology,
https://doi.org/10.1007/978-3-031-44927-7_5

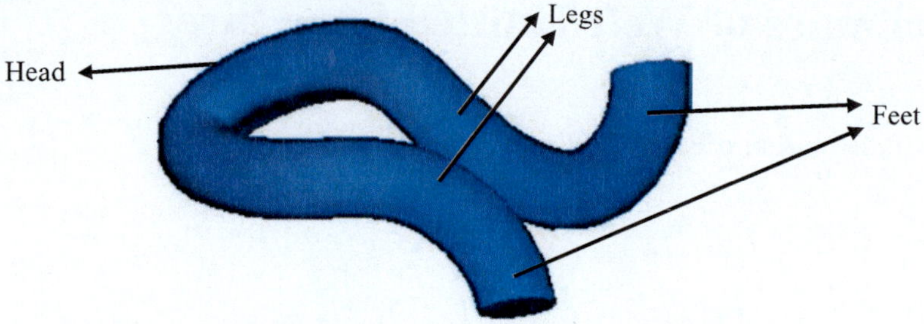

Fig. 1 3D shape of a loop

Fig. 2 Knitted structure with its dimensions

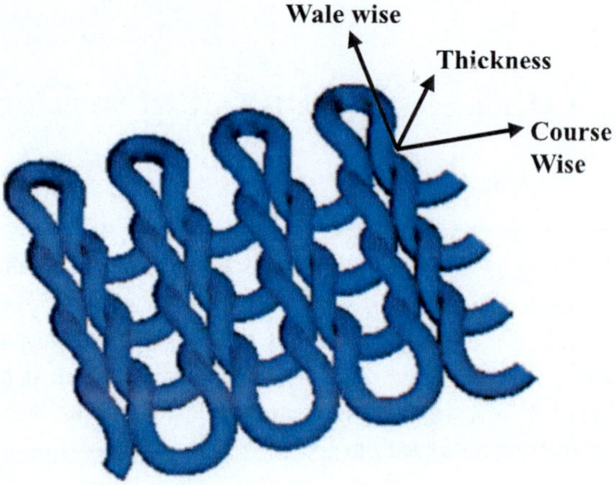

more extensible than woven fabrics, but this is not always true. Later in this chapter, the mechanics of high-strength-knitted fabric is also discussed.

Loop is the basic unit of knitting. It has two legs, one head and two feet as shown in Fig. 1. The legs appear on the front side of the fabric and the head and feet are visible on the back side of the fabric as shown in Fig. 2. When one loop is drawn through another loop, a stitch is formed. A straight row of loops formed consecutively by one yarn is known as course. When many courses are knitted, a vertical row of loops is formed, which is known as wale. A stitch in weft knitting has a three-dimensional (3D) shape. Unlike woven fabrics, yarn in knitted fabric is not straight. Due to this, the thickness of knitted fabrics is not only two times of its yarn diameter as shown in Figure 2. Depending upon the tightness of the knitted fabric its thickness can be greater than 3 times of its yarn diameter.

1.2 Why Knitted Fabrics Are Stretchable?

The answer to this question lies in knitted structure. Knitting is interloping of yarn and when force is applied on course direction, loops are deformed and become straight. Similarly, when force is applied in its wale direction, the spaces between

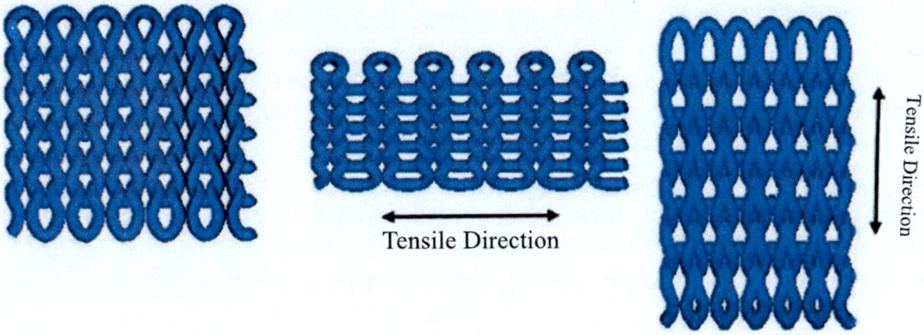

Fig. 3 Behavior of knitted structure when tensile force is applied on it

two wales decrease and fabric stretches walewise, but its dimensions decrease in the other direction as shown in Fig. 3. The more the length of yarn used to make a loop, the more the fabric will be stretchable and vice versa. When force is applied on one axis, the fabric stretches in that direction, but the dimensions of the fabric in other directions are decreased.

Tensile strength and strain of weft-knitted fabrics increase with increasing stitch density of knit as compared with tuck and miss stitch. This result is verified by replacing the knit stitch with tuck of single pique and cross-miss interlock structure [1]. Bending properties such as bending rigidity and hysteresis effect also increase by increasing the knit stitch density. Strain or elongation of knit stitch is greater when compared with knit and miss stitch structure elongation. The compression value of fabrics with a greater number of knit stitch is generally decreased. A structure with more curved stitches generally leads to decreased compression resilience and compression energy due to decreased stitch spaces [1]. The softness and smoothness of a knit structure increase with knit stitches.

Raw materials and structure have a significant influence on the mechanical properties of knitted fabrics. Fabric behavior is affected by testing the direction of the fabric i.e course wise or wales wise. Yarn friction coefficient also affects the deformation process, and yarn with greater friction reduces the strain with reference to load, which is more greater in rib fabrics [2].

Yarn curvature is redistributed more coursewise than walewise, which results in having more elongation of knitted fabrics coursewise. The unstressed shape of knitted structure yields a very low value of Young's modulus as large deformation will be observed when load is applied. Biaxially stressed structure has a high Young's modulus, and stress–strain curve is more linear [3].

The shrinkage behavior of jersey fabric is different from that of rib fabric. The relaxation behavior of jersey fabrics exhibits isotropic dimension change. The lengthwise and widthwise shrinkage always occurs, but rib structures exhibit anistoric shrinkage behavior. The lengthwise shrinkage occurs, with fabric expansion in widthwise or coursewise direction. This behavior can be explained by the planar aspect of the structure. Loops of rib structure tend to lie in or close to the fabric plane, with temporarily distorted dry-relaxed loop shape, elongated by machine-imposed forces, and may change to its relaxed shape retaining an approximate two-dimensional (2D) form. As width expands, length contracts. Once an adequate loop configuration is reached, no further change in dimension will be observed [4].

An idealized knitted structure is made of frictionless, incompressible and uniform load distributed along its straight yarn edges. It is shown that uniaxial and biaxial loading properties of idealized knitted fabric are dependent on the jammed knitted structure. Initially, it has a low stress–strain curve. A sudden change in stress–strain curve of knitted fabric is observed, which shows that all the jamming in the structure is removed, and material behavior starts [5].

A variety of yarn and fabric parameters effect the snagging behaviour of knitted fabrics. It was found that if snag occurs, it causes an increase in tension in snagged yarn. Tension buildup in yarn is influenced by kinetic friction than static friction during snagging [6].

The mechanical behavior of knitted structure cannot be described by only using parameters of materials or its structures. It is the combination of fabric structure, yarn and fiber properties. For a correct explanation of its behavior, explanation of the combination of parameters of structures and material is necessary. When the tensile behavior of knitted structure is observed, it is highly nonlinear. When the stress–strain curve of knitted structure is observed, it can clearly be identified that it is a combination of two parts. The first part is nonlinear and the second is linear. M. De Araujo [7] divides the mechanical behavior of knitted structure into two parts as shown in Fig. 4. The initial mechanical behavior is due to its structure and the lateral behavior is related to the material used. The behavior of knitted structure is

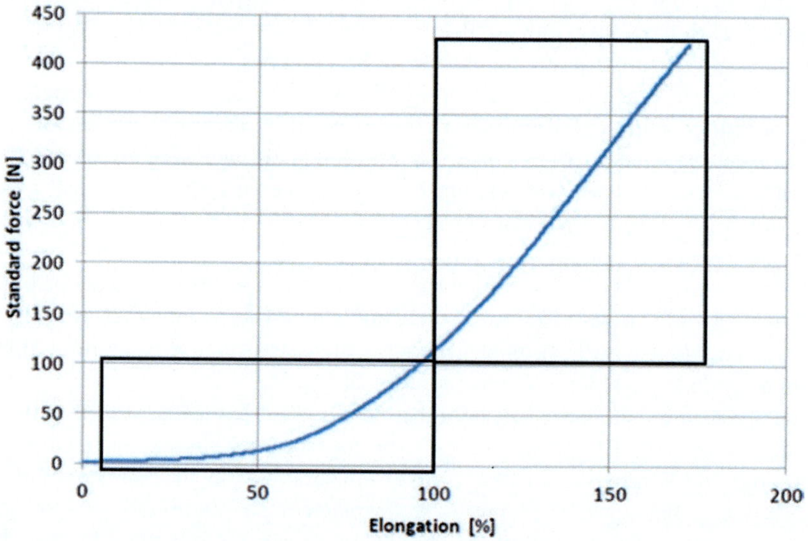

Fig. 4 Stress–strain curve of a knitted structure

more nonlinear than the behavior of materials in stress–strain curve. Elongation observed in the first part is more when compared with the second. This elongation is due to the straightening of the knitted loop. As the second part shows the behavior of materials, the curve is also more linear. Different knitted structures show different mechanical behaviors in the first part of stress–strain curve. Regardless of the material used in knitted structure, the first part elongation will always be present in the structure. It is not possible to theoretically explain Young's modulus of knitted structure because the deformation is nonlinear. As more deformation is observed in the beginning, deformation starts decreasing as force increases. To compare the mechanical behavior of knitted fabrics, Young's modulus stiffness or tensile rigidity can be used. Stiffness is related to flexural rigidity of the yarn (G_0) that is used for knitting. As stitch length is increased in a knitted loop, the radius of curvature (R) in which the yarn bends to form loop shape is also increased. But the bending moment (B) or bending angle of yarn is decreased when the loop length of knitted yarn is increased in order to keep the flexural rigidity of yarn constant. For this case, the yarn parameters may be approximated by using standard numerical calculations. For better performance of knitted structures, one needs to select the appropriate structure along with the material.

$$G_0 = B \times R = \pi d^4 E / 64$$

where

d = yarn diameter;
E = Young's modulus of yarn material.

In the second step, the load is directly applied on materials, i.e., the yarn and second-step deformation are more related to the type of materials used. As load increases, the yarn becomes more compact, and at this stage, there is a minor effect on the structure; thus, a linear, not fully reversible behavior is observed, which is discussed later. In normal tensile behavior, the linearity in stress–strain curve is reversible.

For describing the mechanics of knitted structures, it is necessary to divide them into three main categories:

- Single jersey.
- Rib.
- Interlock.

1.3 Mechanics of Single Jersey Fabrics

Single jersey fabrics are formed by a single cylinder or single-bed machines They have different appearance on the front and back side [8]. On the front side of the fabric, V-shaped legs of loops are visible, whereas on the back side, feet and head of

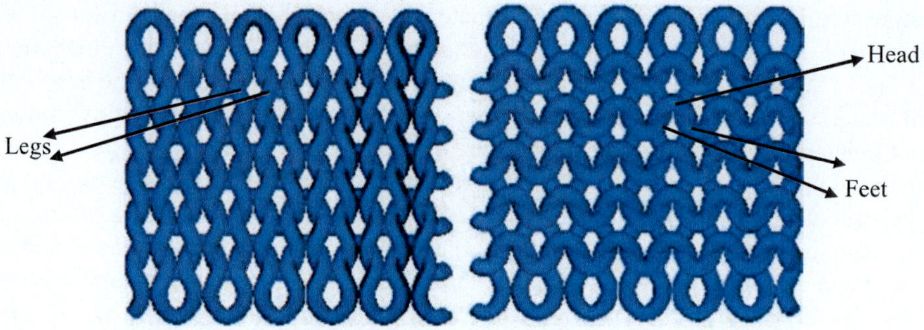

Fig. 5 Single jersey fabric – front and back side

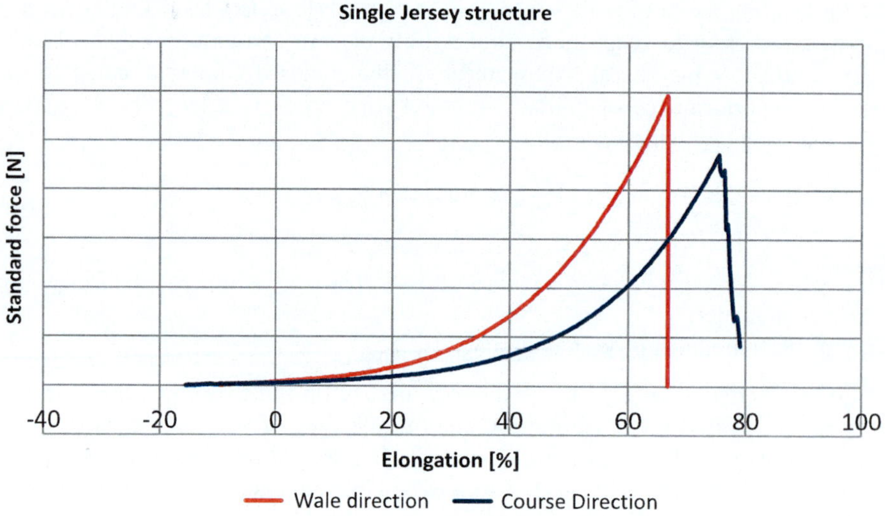

Fig. 6 Stress–strain curve of single jersey structure in course and wale directions

loops are visible as shown in Fig. 5. Single jersey fabrics are not as stable as other fabrics produced from double-bed knitting machines. When knitted on a machine, the single jersey fabric curls in the course direction. The reason for this effect is that its structure is not balanced coursewise and walewise.

When a tensile force is applied on the single jersey structure, its course direction and its loops are deformed and become straight due to which elongation is produced, as discussed earlier. Similarly, when tensile load is applied in the wale direction, spaces between two wales decrease and the length of legs of loops increases, and elongation is produced as shown in figure. The stress–strain curve of plain single jersey structure is shown in Fig. 6. Elongation in course direction is greater than the elongation in wale direction with less value of force in course direction. Although elongation and load values in single jersey structure depend upon the actual stitch length used, but a similar behavior in wale and course directions is always observed. This trend is valid only in single jersey structure having all knit stitches only.

If different types of stitches other than knit stitch are used in a single jersey derivative structure, extension and load values in course and wale directions will vary, and the fabric will have a different mechanical behavior. If tuck or miss stitches are used in a course, the course length will decrease as the amount of yarn required to form a miss or tuck stitch is less than the yarn required for knit stitch. It will cause a decrease in the extension in course direction, and more value of load will be required to produce the extension. If analysis of walewise behavior is done, extension walewise also decreases. When tuck or miss stitch is used in a course due to having less yarn, it decreases wale spacing, which results in producing less extension. But coursewise extension is more effected by using miss or tuck stitches in a course. The stress–strain curve of single jersey derivative is given in Fig. 7. A similar behavior in course and wale at the start of stress–strain curve of single jersey derivative structure is observed as a greater reduction in coursewise extension is observed. Yarn contributes more to tensile strength, and more value of load is observed.

When the tensile behavior of plain single jersey having knit stitches and its derivative having tuck stitches in course is compared, it shows that extension in both course and wale decreases significantly. This is not true that course extension in single jersey fabrics will always be higher than wale extension. If more miss stitches are introduced with appropriate value of stitch length, it is possible that walewise

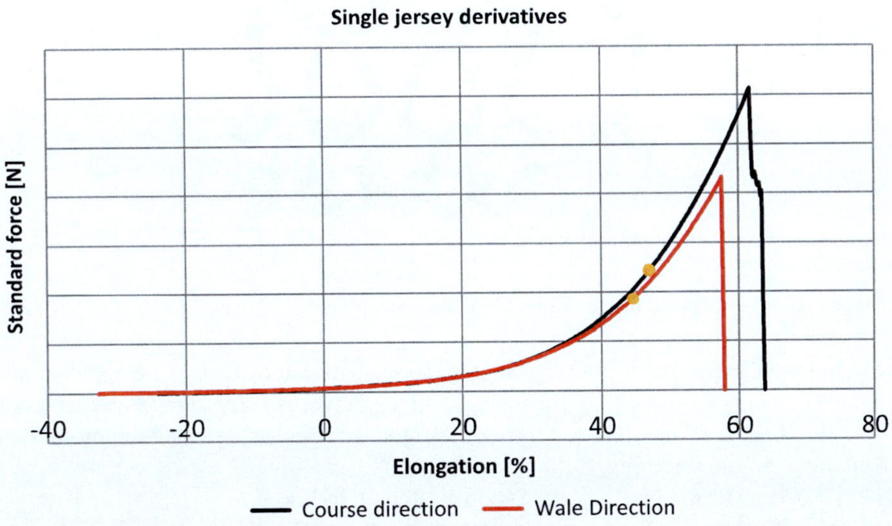

Fig. 7 Stress–strain curve of single jersey derivatives in course and wale directions

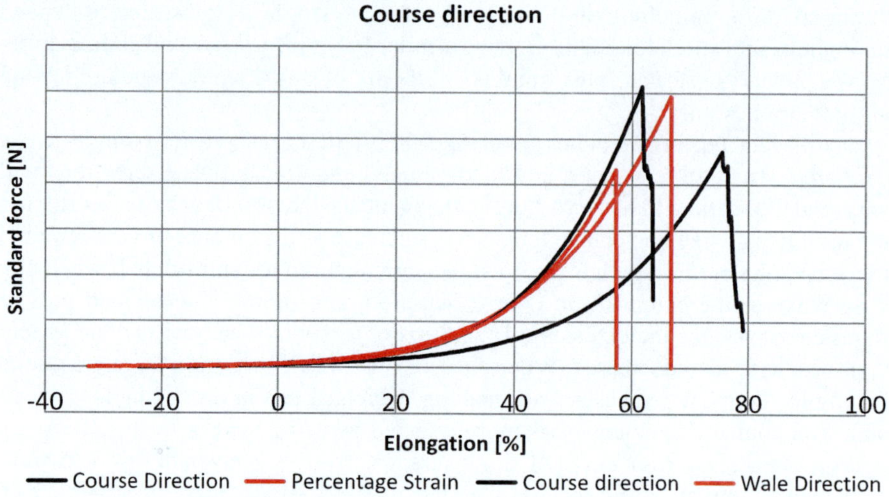

Fig. 8 Combined behavior of single jersey and its derivative in course and wale directions

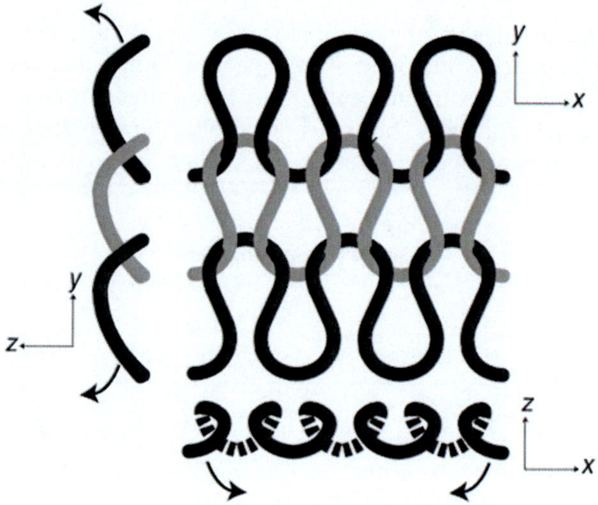

Fig. 9 3D view of knitted fabric. (From M-A. BUENO, p. 88)

extension is higher than coursewise extension. It is interesting to note that more value of load is required to deform single jersey derivative structure from its parent structure in course direction as miss or tuck stitches resist its deformation more than knit stitch as shown in Fig. 8.

A 3D view of the knitted fabric loops is shown in Fig. 9.

1.4 Mechanics of Rib Fabrics

Rib fabrics are manufactured on double-bed knitting machines. These machines can be circular or flat. The two beds of the machine can be dial and cylinder or both flatbeds positioned at different angles. These beds knit the two sides, i.e., front and back side of the fabric. The needles of both beds are alternate to each other due to which both beds can knit simultaneously. A single set of yarn is required to knit the complete course of rib fabric. Depending upon design on each bed that may be same or different, rib fabrics may have same or different appearance on each side of the fabric [8]. Front and back side wales of rib structure are next to each other as shown in Fig. 10, and on both sides of the rib fabric, only legs of the loops appear. Between two wales on the front side, there is another wale on the back side and between two back wales, there is a front wale as shown in Fig. 11. Depending upon the design of knitted fabric, feet and leg of loop may be visible if some needles do not knit. But

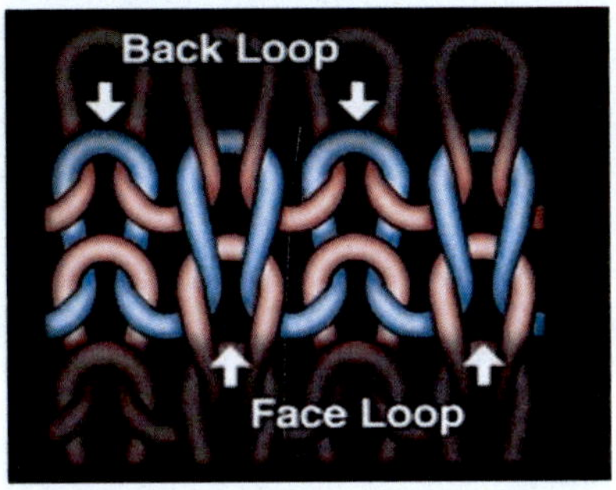

Fig. 10 Face and back loop of rib structure

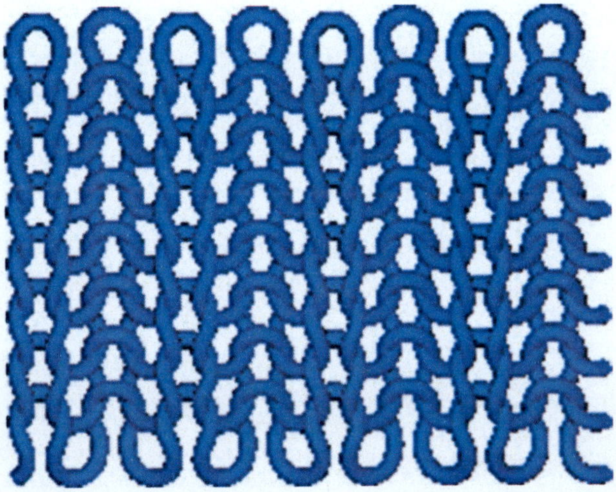

Fig. 11 Rib-knitted structure

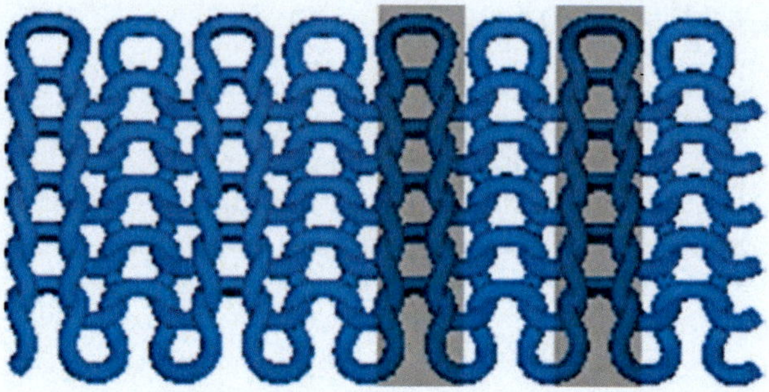

Fig. 12 Appearance of front and back side wale of a rib structure

Fig. 13 Behavior of knitted structure after the application of load

when force is applied to stretch a knitted fabric, space is created between its two wales and feet, and leg of the other side loop also becomes visible. Rib fabrics have the highest stretch ability up to 200%; the reason for this stretch ability is discussed later in this section.

Tensile behavior of rib fabrics is different from single jersey fabrics. When a tensile force is applied along its course direction, elongation is produced, which is two times greater than the single jersey fabrics. This greater elongation is related to its structure. Rib fabrics have stitches double to single jersey fabrics as loops are knitted on both sides of the fabric. Rib fabrics have two sets of wales or loops, i.e., front and back side. When tensile force is applied in course direction, both front and back side loops will be straight and a double elongation will be produced with the same amount of force than the elongation produced from a single set of front side loops of single jersey fabrics as shown in Fig. 12. In the direction where force is applied, specimen dimension increases, but it decreases in the other direction as shown in Fig. 13.

From the stress–strain curve of rib structure shown in Fig. 14, it can be clearly identified that a great amount of elongation is produced in course direction than the wale direction. The mechanism of this elongation is already discussed earlier.

1.5 Mechanics of Interlock Fabrics

Interlock fabrics are also knitted on double-bed circular knitting machines. Unlike rib machine, only circular interlock machines are only available with cylinder and dial. Flatbed interlock machines rarely exist. Like rib-knitted fabrics, in interlock

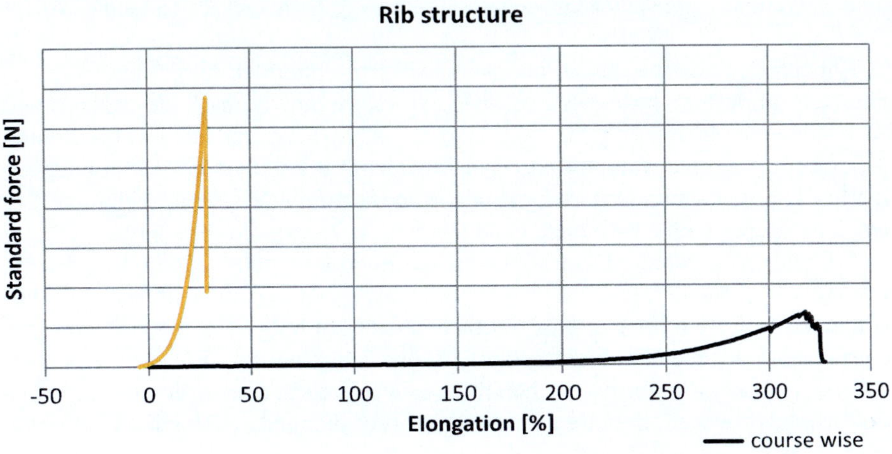

Fig. 14 Stress–strain curve of rib structure

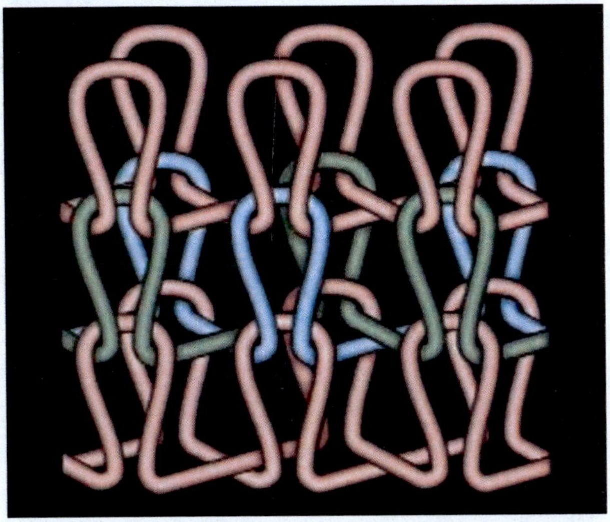

Fig. 15 Interlock course formation from two feed presented in different colors

fabric, two different beds knit the two different sides of interlock fabrics. In interlock fabrics, the position of cylinder and dial needles is directly opposite to each other due to which opposite needles cannot knit at the same time. Unlike rib fabrics, two sets of yarns are required in interlock machine to complete one course as shown in Fig. 15. It is not possible in interlock fabrics to knit the opposite loops at the same time. When needles of dial or cylinder knit, the opposite needle of cylinder or dial remains at rest. In this way, alternate needles of dial and cylinder knit in the first feed, and remaining needles knit in the second feed. In this way, two feed complete one course of interlock fabrics.

Depending upon the design on each side of the interlock fabric, both sides of interlock fabric may have same or different appearance on each side [8]. But on each side of the fabric, only legs of loops are visible. Front and back wales of interlock fabrics are directly opposite to each other as shown in the top view of interlock fabrics. Interlock structures are more stable and require more force for deformation when compared with single jersey and rib fabrics. To explain the stability of interlock fabrics, it is required to understand the interlock structure, which is formed from two feed. In interlock fabrics, one yarn makes only half stitch of a course when compared with rib fabrics as only alternate needles of both beds can knit, and yarn consumed in one feed in interlock fabric is also less. Now, when force is applied to deform the interlock fabric, only half stitches will deform a feed, and less deformation will be produced. In other words, the loops of dial and cylinder are opposite to each other and when they deform after application of force, elongation is produced by opposite loops. But when force is applied to deform two side loops directly opposite to each other, the force required must be greater than the single side loops in jersey fabrics. Further, yarn friction between two sets of yarn in one course in interlock fabrics also plays a role during deformation. The representation of interlock structure in different directions is given in Fig. 16.

The interlock course formation from two feeds is shown in Fig. 17.

From the stress–strain curve of plain interlock structure given in Fig. 18, like single jersey structure, it can be observed that greater elongation is present in course

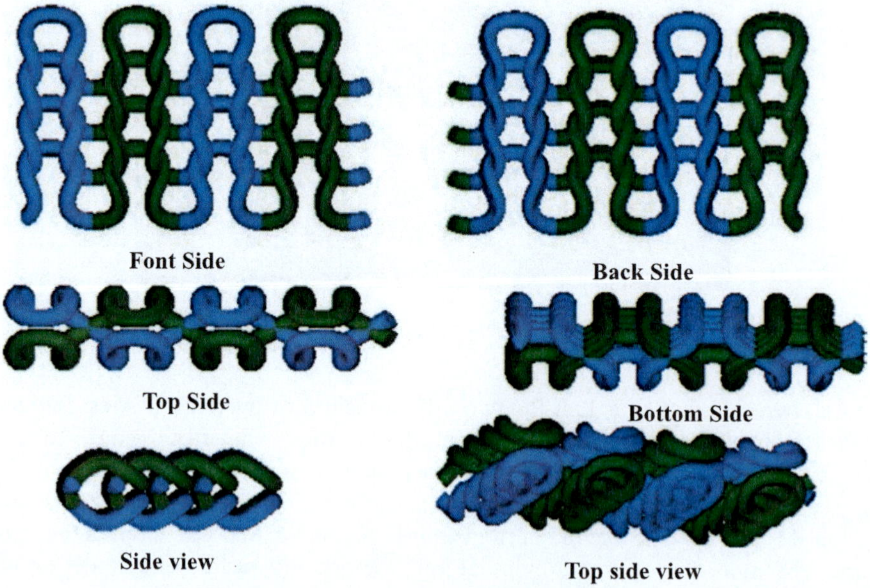

Fig. 16 Representation of interlock structure in different directions

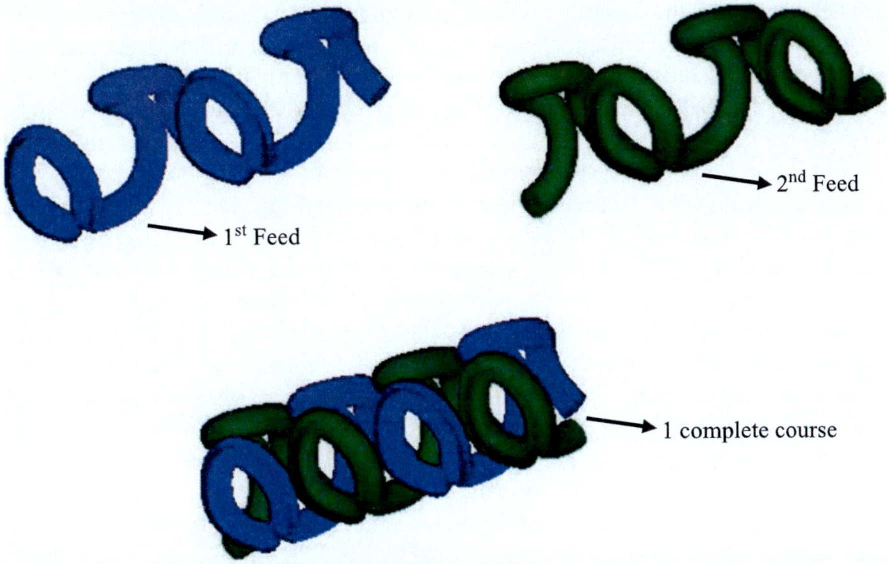

Fig. 17 Formation of an interlock course from two feed

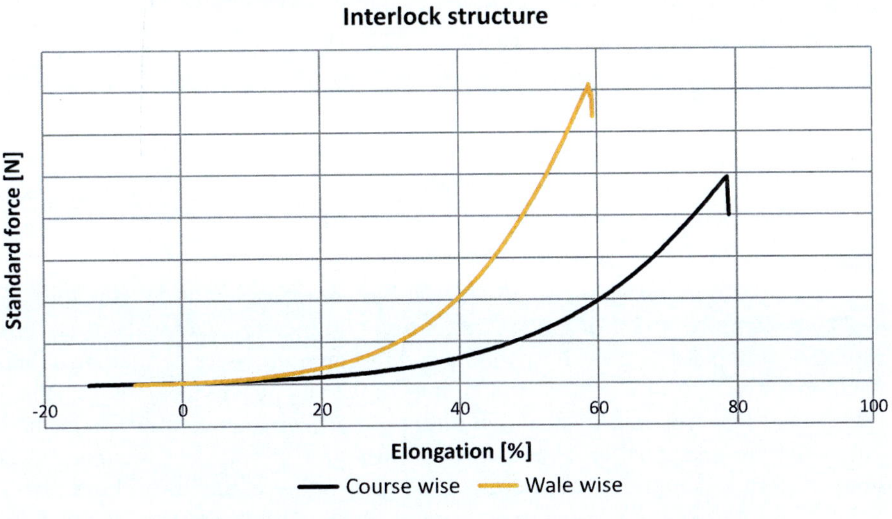

Fig. 18 Stress–strain curve of interlock structure in course and wale directions

direction. Interlock structures have two sets of front and back side loops, but unlike rib structure when tensile force is applied, the elongation produced is of one set of loops only. As front and back side loops are directly opposite to each other, the deformation produced is of one loop only. Thus, the interlock structure is considered to be more stable as less deformation is produced. Walewise extension of interlock structures is less than the coursewise extension. But coursewise extension can be decreased by using a combination of miss and tuck stitch in interlock derivative structures. Yarn of a feed in interlock structure is interlocked in the front and back side of the fabric due to which when tensile force is applied in wale direction, the wales spacing does not decrease for producing walewise extension like in single jersey fabrics. Owing to the use of two feed in a course, less space is available for the yarn to produce walewise extension.

Nyi Nyi Htoo [9] constructed the three-dimensional model of weft-knitted loop by using simulation yarns. The yarn used for simulation was constructed from cross section consisting of mass points connected with spring within the mass points. Along all the cross sections, three coordinates x, y and z were allocated. Spring was connected to express the twist in the simulated yarns as given in Fig. 20 between the adjacent sections with inclined angle. Yarn model was then transformed into a three-dimensional loop model, which was created by modifying the Pierce model, and a new model of knitted loop was developed.

When load is applied in course direction in developed model, it changes its dimensions of course and wales. As this is observed physically and in the developed model, wales spacing increases, and courses are compressed. Similarly, when tensile load is applied in wale direction, model changes in both wale and course directions. But in this case, contradictory to load applied in course direction, the wales are extended and courses are compressed. The deformation due to tensile in course and wale directions and newly deformed position can be described by the below relation:

$$B_1 = B \times \frac{OB}{OC}$$

$$B_1 = B \times \frac{h}{H}$$

where

B is the position of section B, OB/OC and h/H are referred to as length ratio.

The bending behavior of knitted structure in both course and wale directions was also stimulated. Firstly, loop model and its curvature were defined, and then the model was bent to a certain amount in positive and negative curvature. When it was bent in a positive or negative direction, all sections of model and their position were changed. For bending loop structure in wale and course directions, different curvatures were considered as bending behavior on both directions is not same. Secondly, the position of all curvatures was defined by formulae, and after bending in course and wale directions, different positions were calculated using formulae. The different formulae used for calculating the position are given below:

$$x = \frac{1-\cos K}{K}$$

$$z = \frac{\sin K}{K}$$

$$x = \frac{1-\cos K}{K}$$

$$y = \frac{\sin K}{K}$$

$$x' = \frac{1-\cos K'}{K'}$$

$$z' = \frac{\sin K'}{K'}$$

$$x' = \frac{1-\cos K'}{K'}$$

$$y' = \frac{\sin K'}{K'}$$

where

K is the curvature of edge sections, K' is the curvature of non-edge sections.

For course and wale bending, below formulae are used:

$$K' = K \times \frac{l}{L}$$

$$K' = K \times \frac{h}{H}$$

where l is the length or position between one edge of mode and non-edge of section and L is the length or position between two edges of model in course direction. Similarly, h is the length or position from bottom to non-edge and H is the length from top to bottom in a simulated model. The simulated images of bending and tensile are given in Fig. 19. The image of mass point along the yarn axis and transformation of yarn model into loop is shown in Fig. 20.

A modified Pierce model is shown in Fig. 21, and the deformed loop structure under load in course and wale directions is represented in Fig. 22.

G. A. Carnaby studied the shear properties of wool-knitted structures. He found that shear resistance of wool-knitted fabrics is low as compared to other woven and non-woven wool products. It was observed that knitted fabrics had a low value of secondary modulus. A large hysteresis effect was found in shear properties of knitted fabrics. The shearing behavior of rib fabrics was found to be more anisotropic than interlock fabrics when shared in course and wale directions. The reason behind anistropy is the different elongation behavior of rib fabrics [10].

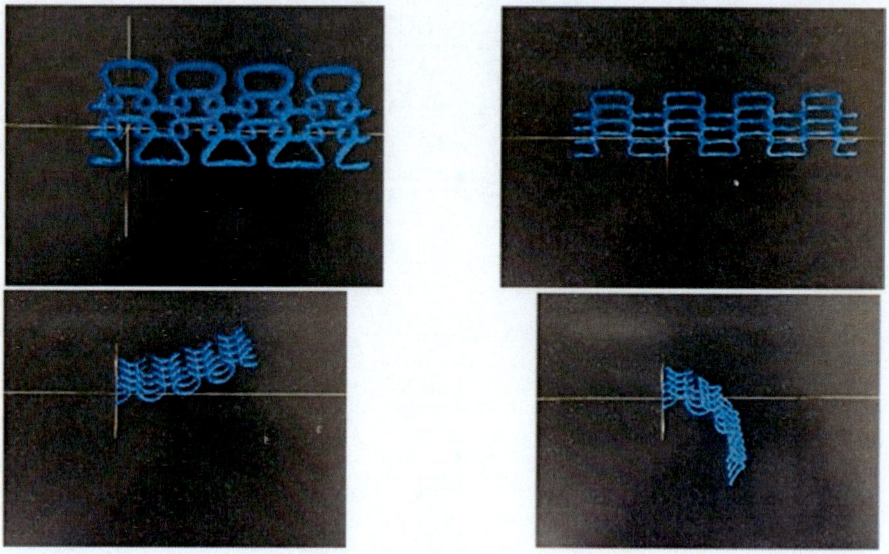

Fig. 19 Images of simulated model during bending and tensile loading in course direction. (From Nyi Nyi Htoo, 2017)

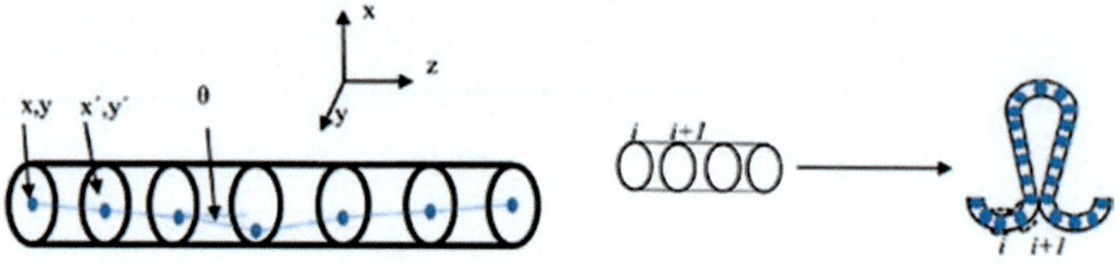

Fig. 20 Image of mass point along yarn axis and transformation of yarn model into loop. (From Nyi Nyi Htoo, 2017)

Fig. 21 Modified Pierce model. (From Nyi Nyi Htoo, 2017)

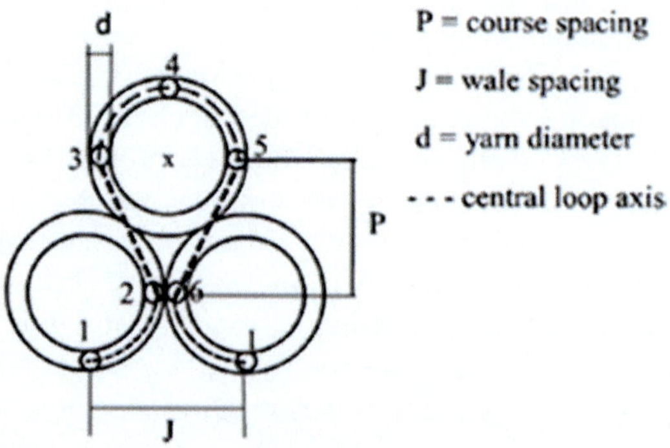

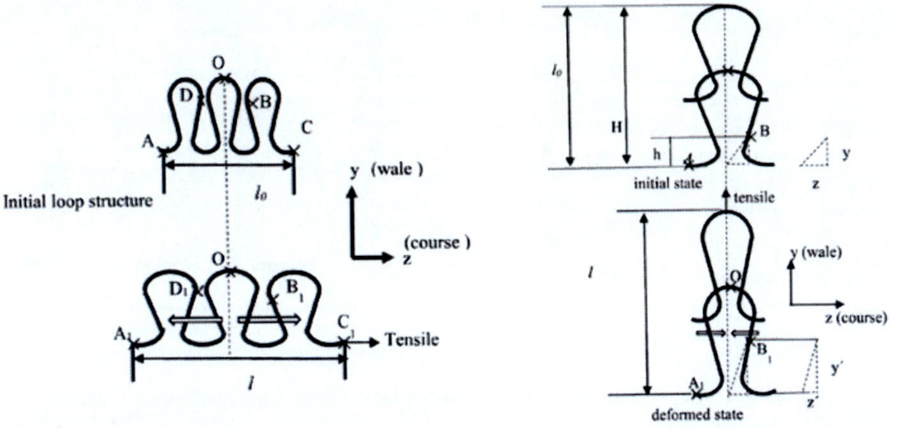

Fig. 22 Deformed loop structure under load in course and wale directions. (From Nyi Nyi Htoo, 2017)

Mark S. Yeoman modelled the non-linear behavior of knitted fabrics. The aim of his study was to provide a tool to model knitted fabrics for medical application. The important properties that need to be modelled are its strain with reference to applied force. He found that nonlinear strain produced in knitted fabrics can be used to describe the mechanical behavior of knitted fabrics [11].

M Szabo studied the change in dimensions of single jersey fabrics after knitting. Tubular fabrics were knitted using 58 Nm yarn density of 100% cotton yarn. Fabrics were relaxed for 72 h after knitting in standard conditions. Wale density was found to be the fundamental parameter for dimensional change of jersey fabrics. If an optimum level of wale density is used or found, then other factors such as tension, feed rate and speed will have less influence on the dimensional change of knitted fabrics [12].

R. J. Hamilton studied the bending and recovery properties of knitted fabrics. He found the bending process of knitted fabric very complex, and it required several parameters to explain hysteresis in bending of knitted fabrics. He used fluxural rigidity, hysteresis and curling couple parameters to explain the bending behavior. The magnitude of each parameter was reduced in relaxed fabric [13].

Yordan Kyosev developed a model of loop for knitted fabrics. As loop dimensions changes with time, a model based on dynamic model where the loop geometry is dependent on time was developed. Finite element analysis was used to define the geometry of loop. The initial loop geometry was defined using splines among certain key points of yarns, and then yarn axis and volume were defined. The defined loop geometry has three wales and three courses corresponding to fabric knitted on a circular machine with E32 with 2.34 mm dimension of modelled loop. The modelled fabric was deformed at a rate of 10 mm/s and after only 0.039 s, 17% elongation was produced; 17% and 34% elongation are shown in Fig. 23. A higher value of load was required when yarn friction between loops was considered during deformation. Much higher time was required for the calculation of elongation up to 125%

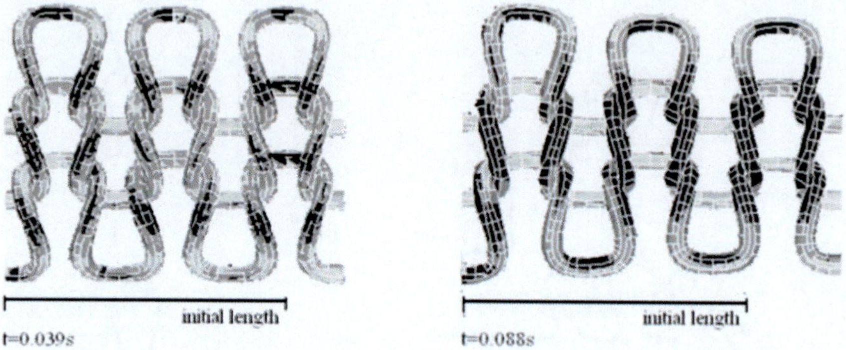

Fig. 23 Deformed loop elongation with reference to time. (Form Yordan KYOSEV 2006)

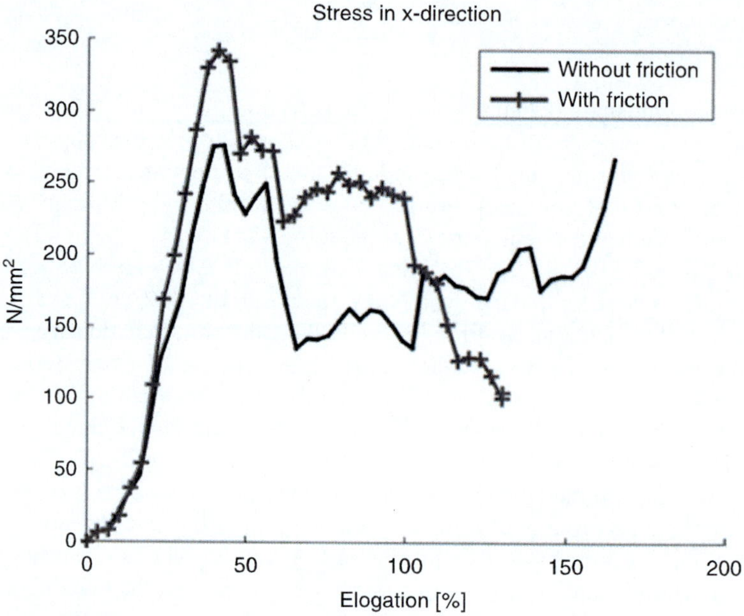

Fig. 24 Stress–strain behavior of stimulated model with and without friction. (From Yordan KYOSEV, 2006)

for this model. There was deviation in the calculated and modelled behavior under tensile force. Smaller unit of mesh can be used for more accuracy as yarn surface becomes more smooth, but it requires a large calculation and a very long calculating time. It was concluded that simulation of load in the first 100% elongation can be modelled. This modelled can be used for research only using high performance computers [14]. The stress–strain behavior of stimulated model with and without friction is presented in Fig. 24.

1.6 Mechanics of Miscellaneous Functional Knitted Structures

Aramid knitted structures are extensively used for ballistic protection. Studies have been conducted to understand the performance of filaments yarns and their knitted fabrics under low velocity impact [15]. The knitted fabrics and their load-extension behavior are shown in Fig. 25.

Knitted structures show excellent shape retention and bagging tendency based on their stretchability. Computer-assisted techniques are based on the mechanics of deformation in the macrostructure of several knitted fabrics [16]. The structural deformation in knitted fabrics can be simulated using computer-assisted design (CAD) as shown in Figs. 26 and 27.

Knitted fabrics when subjected to shear result in micro and macro buckling in the structure. Theoretical studies have been conducted using micro-mechanical models considering the stresses and the deformations under static as well as dynamic conditions [17].

The material geometry and architecture play a significant role in engineering the mechanics and resulting performance of knitted structures. Numerical and computational analysis have been conducted on several knitted geometries. Stress distribution at the cross-over points have been analyzed using finite element analysis [18]. A few models are shown in Fig. 28.

Biaxial knitted fabrics are much more robust and are widely used in construction materials. Several studies are conducted to understand the deformation mechanism under shear loading of such fabrics [19]. The frame test method is used to experimentally determine the biaxial deformation and shear in knitted fabrics as shown in Fig. 29.

Functional knitted fabrics are designed to exhibit nonlinear and gradual deformation based on level of stress and the direction of loading. Such knitted structures are widely used in composites and technical textile applications. The mathematical and numerical models are based on material micro-mechanical properties and geometrical orientation of loops. 2D and 3D constitutive models based on empirical relations are implemented to derive the simulated stress distribution as shown in Fig. 30 [20].

Coated knitted fabric composites show nonlinear mechanical behavior under tensile loading. Fractional constitutive models are employed to determine the stress–strain relationship in such knitted fabric composites [21]. Uniaxial tensile tests reveal the viscoelastic nature of knitted fabric structures as shown in Fig. 31.

Knitted textiles are widely used in biomedical applications, e.g., replacement of bones, muscles, or even cardiovascular valves. The wale and course–coupled nonlinear behavior in medical applications has been studied extensively by analyzing the mechanics in terms of friction, shear, elasticity and flexibility [22]. An example of a mechanistic model is shown in Fig. 32.

Hybrid composites with stitched woven–knitted fabrics were developed with a view to improve the impact performance. The knitted fabrics are placed in the inner layers of eight-layered composite structures to absorb the impact energy [23].

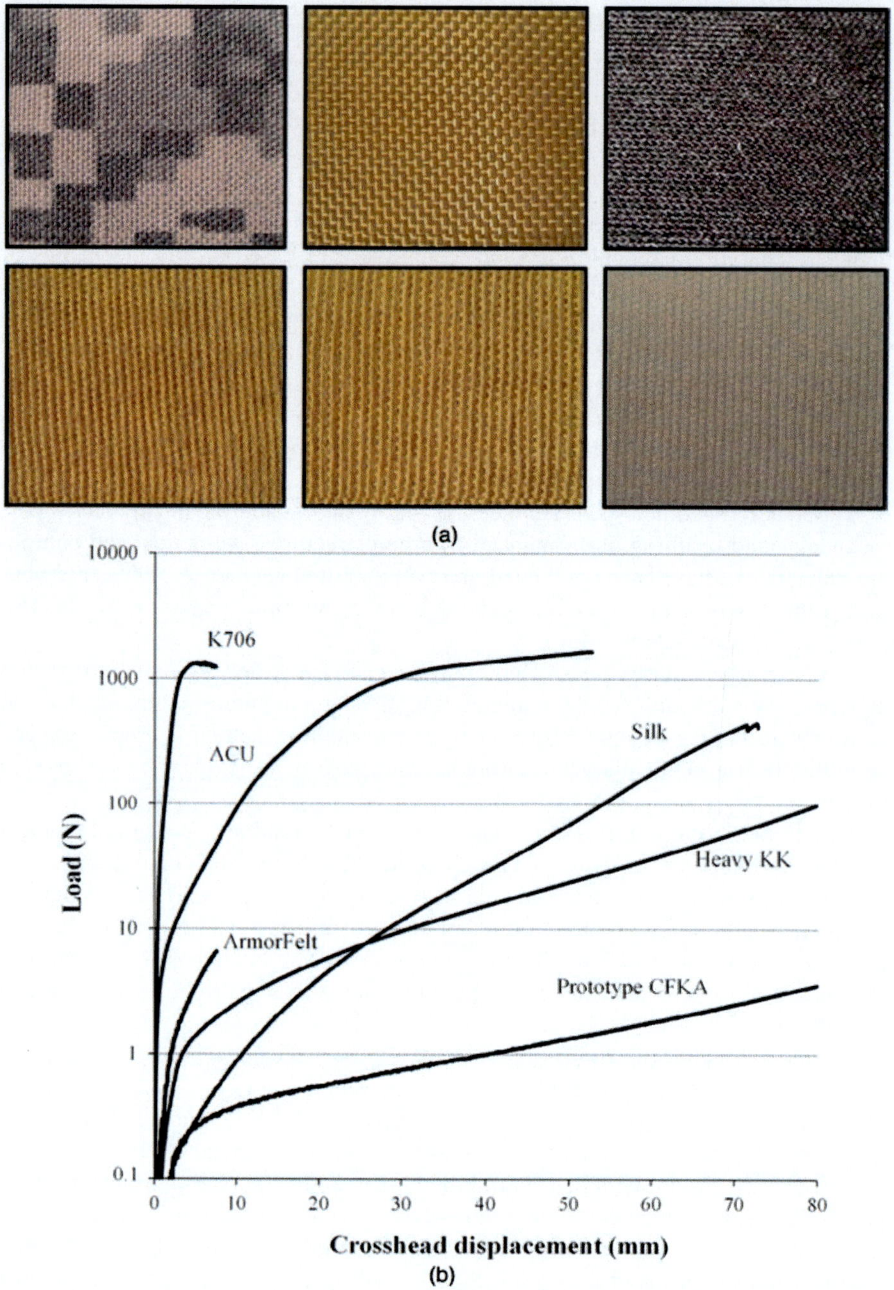

Fig. 25 (**a**) Filament yarn knitted structures and (**b**) load-extension behavior. (A.K. Dwivedi et al./ International Journal of Impact Engineering 96 (2016) 23–34)

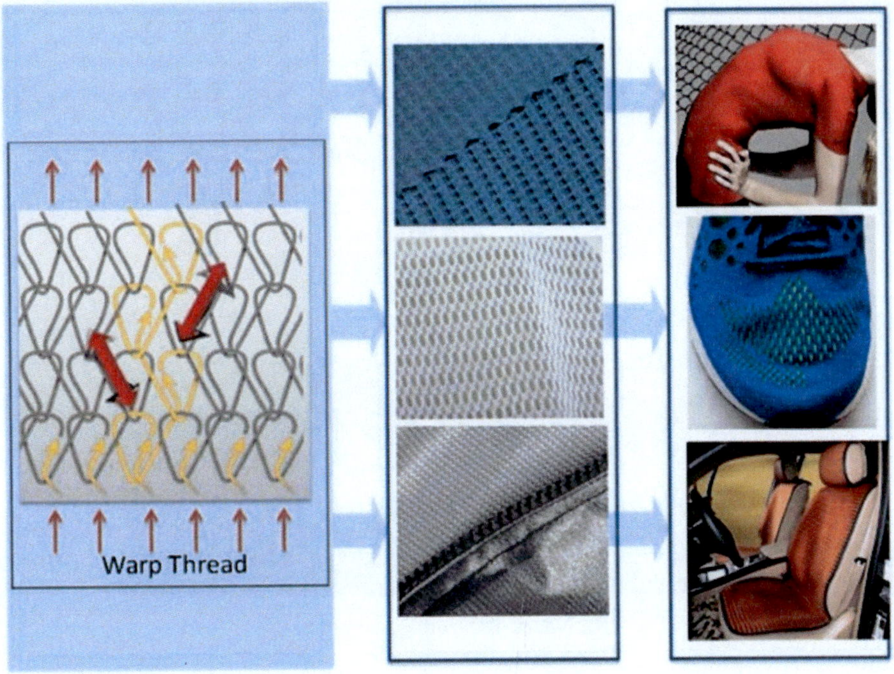

Fig. 26 Bagging tendency of knitted structure and their deformation mechanics. (A. Mao et al./ Computer-Aided Design 75–76 (2016) 61–75)

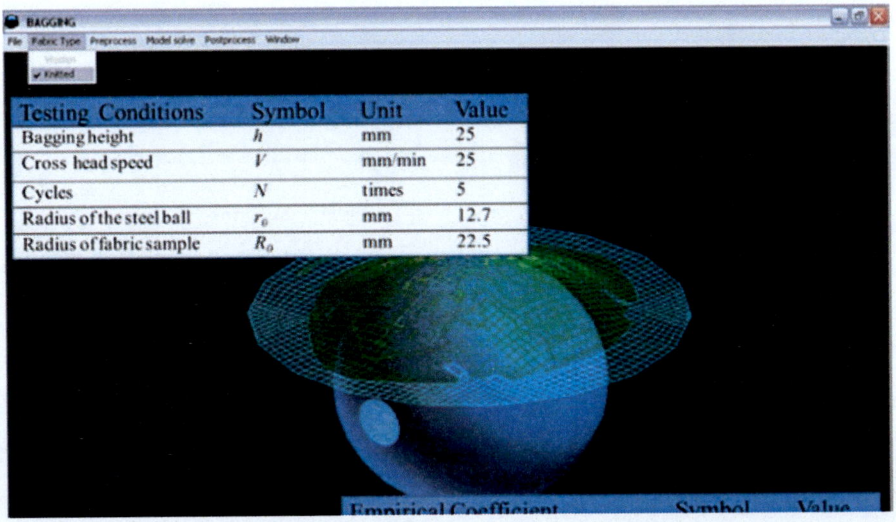

Fig. 27 Computer-assisted design system for bagging deformation in knitted structures. (A. Mao et al./Computer-Aided Design 75–76 (2016) 61–75)

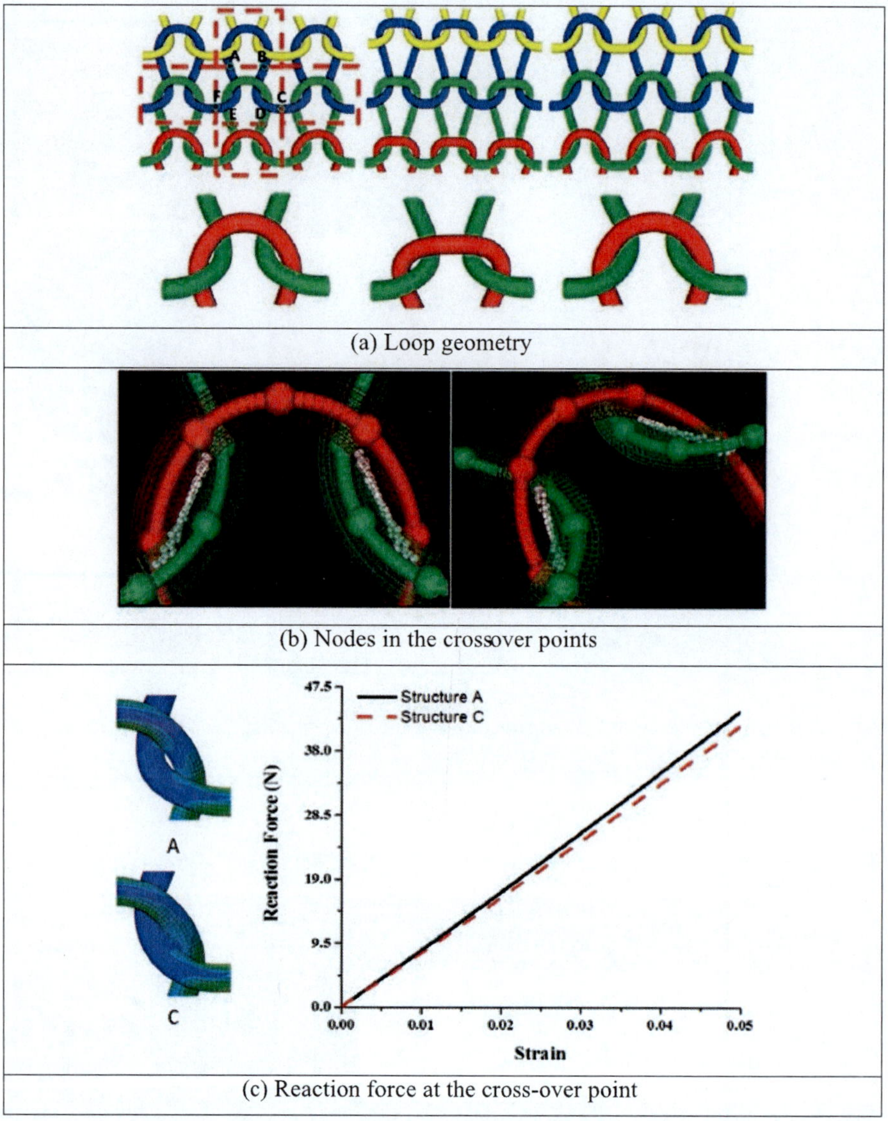

Fig. 28 Micro-mechanical models of knitted architectures. (International Journal of Solids and Structures, Volume 109, 15 March 2017, Pages 101–111)

Mostly rib-knitted fabrics were placed together with 2D woven fabrics for improved impact energy absorbance. An illustration is provided in Fig. 33.

The ballistic impact performance of single jersey knitted fabric has been a topic of major interest. Several experimental and computational studies have been reported regarding the micro-mechanics of the deformation and bulging behavior

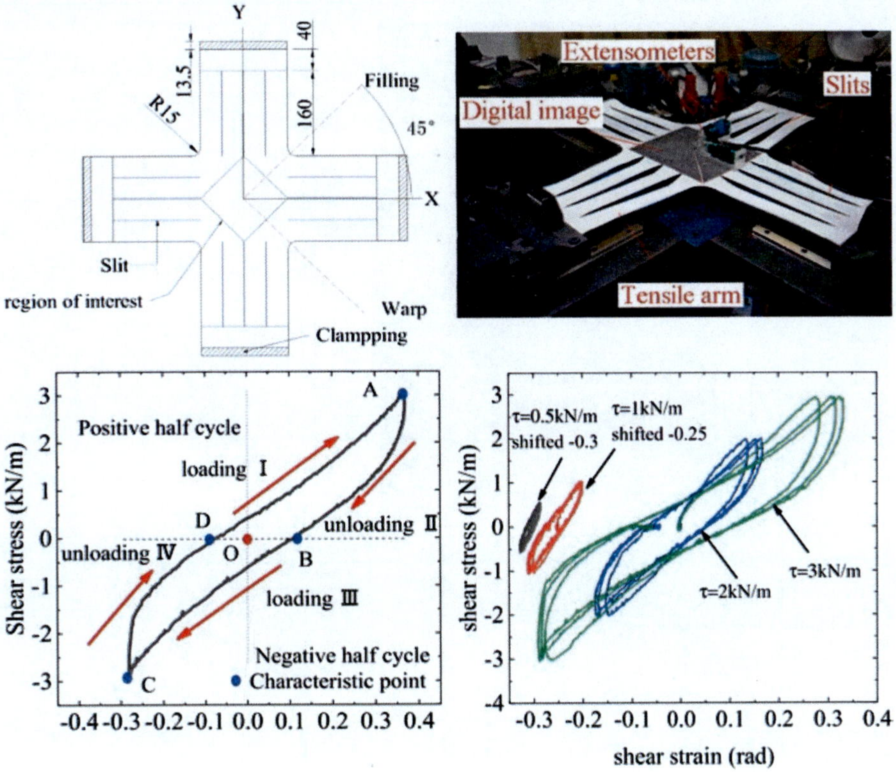

Fig. 29 Biaxial deformation of knitted fabrics. (Construction and Building Materials 304 (2021) 124517)

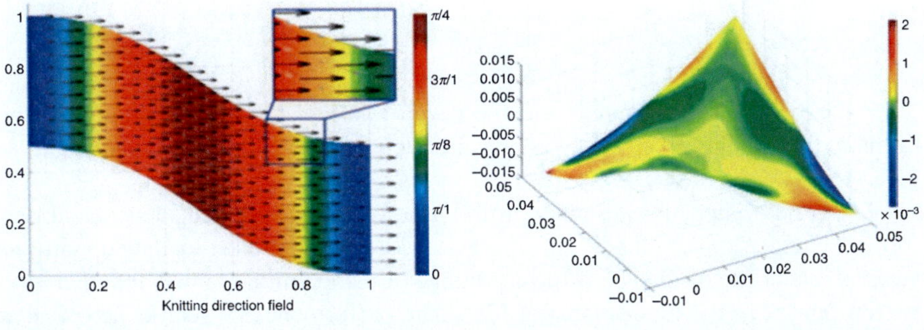

Fig. 30 Micro-mechanical numerical solution of knitted fabric structures. (Composites Part B 202 (2020) 108416)

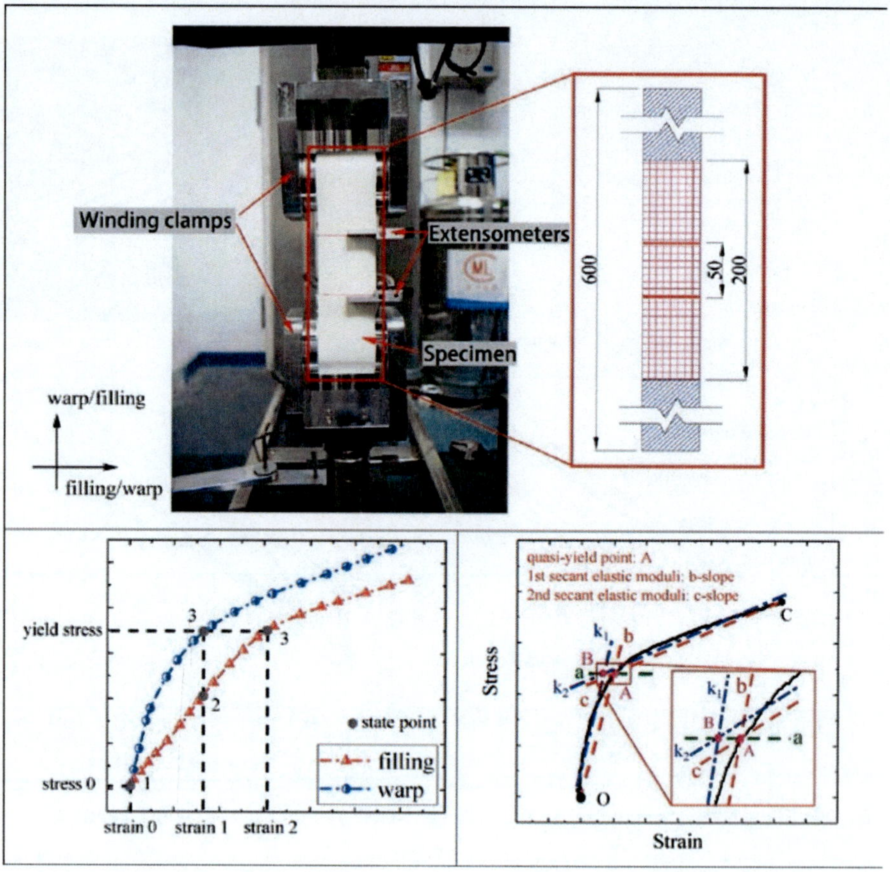

Fig. 31 Viscoelastic and nonlinear behavior of coated knitted fabric composites. (Polymer Testing 107 (2022) 107464)

[24]. The distribution of stress and the corresponding radial deformation can be evaluated with high precision by using finite element analysis as demonstrated in Fig. 34.

3D hybrid composites are made from knitted fabrics, and their mechanical performance under tensile, bending, compression, shear and impact loading conditions are evaluated. The results show the advantage of using knitted fabrics together with woven fabrics in hybrid composites [25]. The performance of such composites at different temperatures are depicted in Fig. 35.

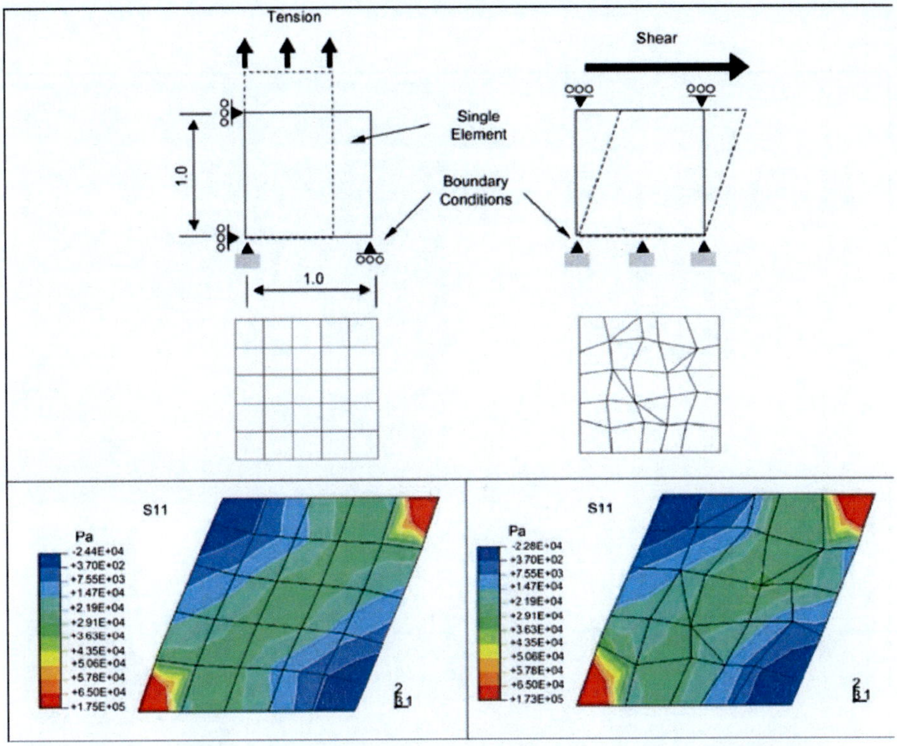

Fig. 32 Mechanistic model of a knitted biomedical textile. (Biomaterials 31 (2010) 8484e8493)

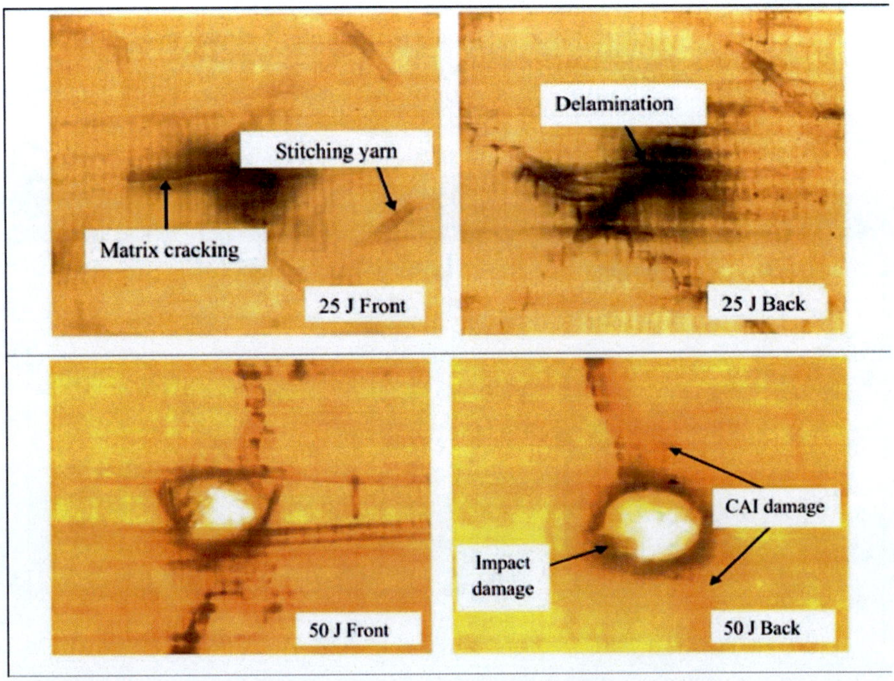

Fig. 33 Impact mechanics of stitched woven–knitted composites. (A. Aktaset al./Composite Structures 116 (2014) 243–253)

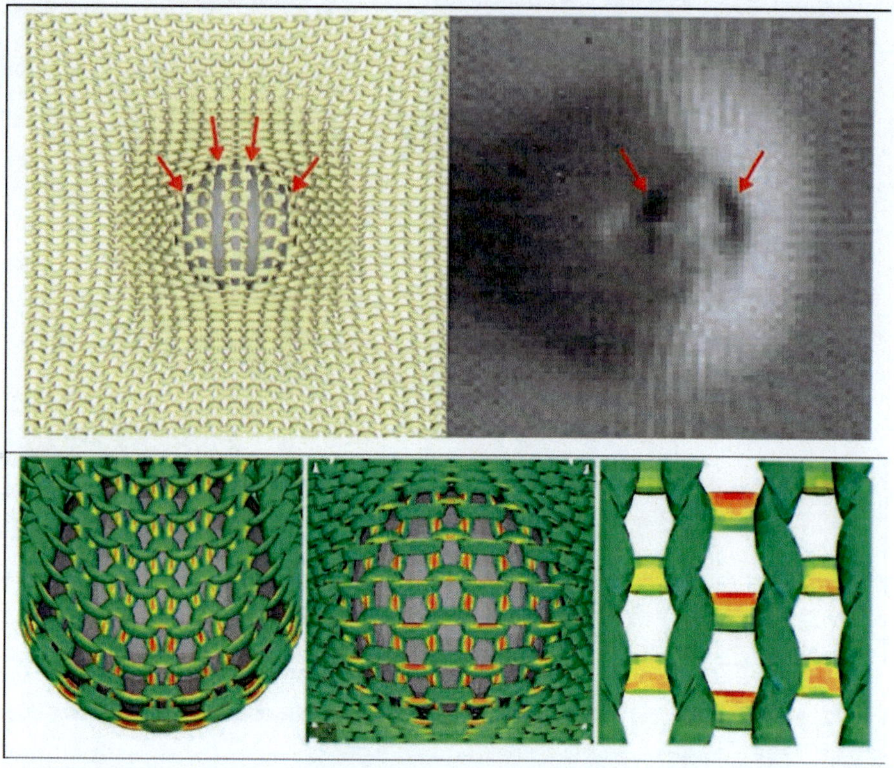

Fig. 34 Finite element analysis of the radial deformation during ballistic impact. (P.J. McKee et al./Composite Structures 162 (2017) 98–107)

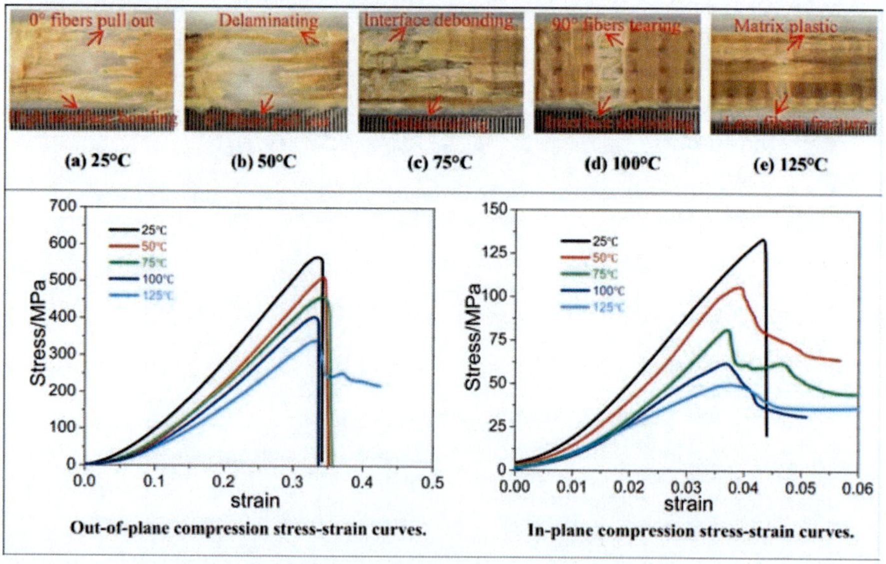

Fig. 35 The mechanical performance of hybrid knitted composites at different temperatures. (Composite Structures 246 (2020) 112340)

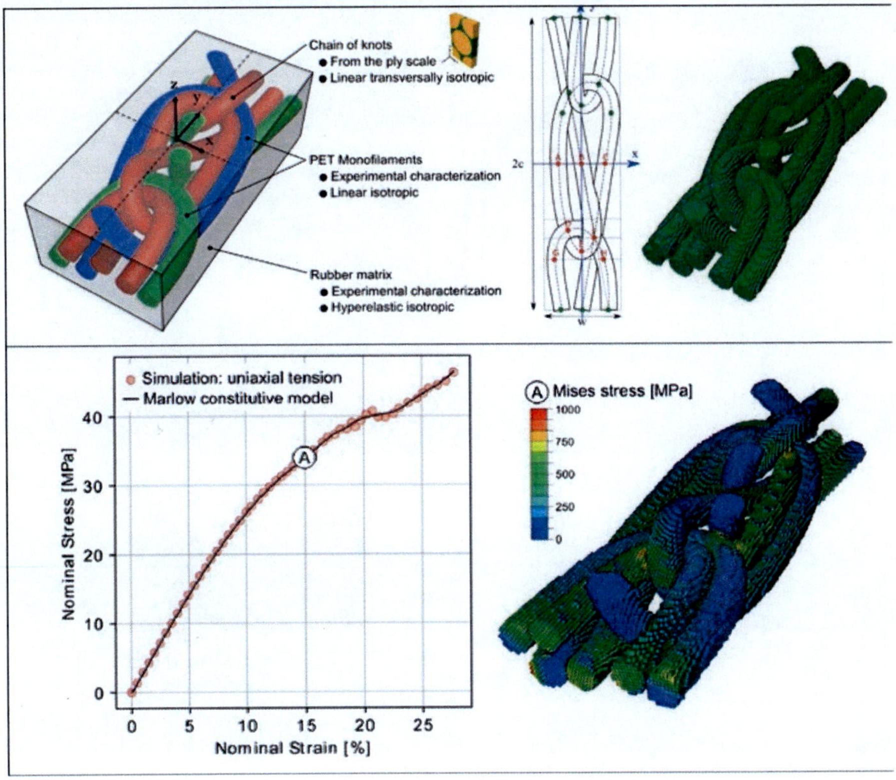

Fig. 36 Multiscale mechanical models for spacer knitted fabrics composites. (Composite Structures 257 (2021) 113139)

Spacer knitted fabrics have several unique characteristics owing to their spatial geometry. Multiscale mechanical models are developed to explain the deformation under compression, tension and bending in spacer knitted fabrics [26]. Composites reinforced with spacer knitted fabrics can exhibit exceptional mechanical behavior as shown in Fig. 36.

Knitted textiles are widely used in biomedical fields. Some permanent set knitted fabrics are used in medical implants, and their anisotropy of performance is characterized by using computational modeling [27]. Uniaxial and biaxial tensile mechanism is used to analyze their performance as an implant in medical fields. Illustrations are given in Fig. 37.

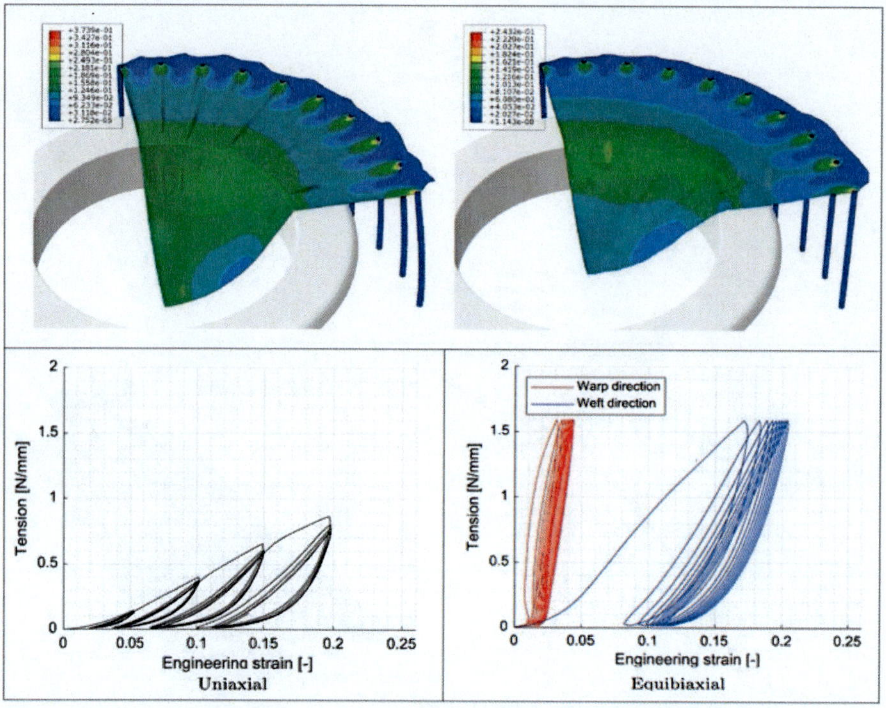

Fig. 37 Uniaxial and biaxial deformation of knitted implants. (Journal of the mechanical behavior of biomedical materials 114 (2021) 104210)

References

1. M. Choi, S. Ashdown, Effect of Changes in Knit Structure and Density on the Mechanical and Hand Properties of Weft-Knitted Fabrics for Outerwear, Text. Res. J. **70**(12), 1033–1045 (2000)
2. L. Ciobanu, Experimental study on the mechanic behaviour of weft knitted fabrics. Fibres Text. East Eur. **2**(29), 34–39 (2012)
3. R. Postle, R. Postle, Structural mechanics of knitted fabrics for apparel and composite materials. Int. J. Cloth. Sci. Technol. **14**(3–4), 257–268 (2002)
4. K.W. Fabrics, Half-Cardigan machine-washing tumble-drying. 854, 1095–1106 (1968)
5. B. Hepworth, 13 – The biaxial load-extension behaviour of a model of plain weft-knitting – Part I. **5000**(June) (2016)
6. N. Carolina, S. Backer, 1 Structures, pp. 262–271
7. M. De Araujo, R. Fangueiro, H. Hu, Weft-knitted structures for industrial applications. Adv. Knitt. Technol., 136–170 (2011)
8. E. Wood, 20. Formation and properties of knitted structures, pp. 1–8 (2009)
9. F. Sci, T. Society, and F. Science, Abstract The three-dimension weft- knitted loop model was constructed using the simulated yarn which was made by mass-spring system. The simulated yarn was constructed with the cross-sections consisting of the mass-points connected by the springs within, **73**(5), 105–113 (2017)

10. G. A. Carnaby, The Journal of The Textile 12 — The SHEAR PROPERTIES OF WOOL WEFT-KNITTED STRUCTURES, March 2015, pp. 37–41
11. M.S. Yeoman, D. Reddy, H.C. Bowles, D. Bezuidenhout, P. Zilla, T. Franz, Biomaterials A constitutive model for the warp-weft coupled non-linear behavior of knitted biomedical textiles. Biomaterials **31**(32), 8484–8493 (2010)
12. S. Vasile, B. Malengier, D. Kopitar, Z. Skenderi, Aspects of the dimensional changes of jersey structures after knitting process Aspects of the dimensional changes of jersey structures after knitting process (2017)
13. W.P. Fabrics, of Wool Plain-Knitted Fabrics, c, pp. 336–343
14. Y. Kyosev, Computational model of loops of a weft knitted fabric, pp. 1–6 (2006)
15. A.K. Dwivedi, M.W. Dalzell, S.A. Fossey, K.A. Slusarski, L.R. Long, E.D. Wetzel, Low velocity ballistic behavior of continuous filament knit aramid. Int. J. Impact Eng. **96**, 23–34 (2016)
16. M. Aihua, L. Jie, L. Yi, L. Yinglei, H. Yanxia, Knitted fabrics design and manufacture: A novel CAD system for qualifying bagging performance based on geometric-mechanical models. Computer-Aided Design **75–76**, 61–75 (2016, June)
17. Y.T. Zhang, C.Y. Liu, R.X. Du, Buckling analysis of plain knitted fabric sheets under simple shear in an arbitrary direction. Int. J. Solids Struct. **44**, 7049–7060 (2007)
18. D. Liu, D. Christe, B. Shakibajahromi, C. Knittel, N. Castaneda, D. Breen, G. Dion, A. Kontsos, On the role of material architecture in the mechanical behavior of knitted textiles. Int. J. Solids Struct. **109**(15), 101–111 (2017, March)
19. J. Chen, Y. Xia, B. Zhao, W. Chen, M. Wang, J. Fan, R. Zhang, Detailed characteristics of shear stiffness for coated biaxial warp-knitted fabrics subjected to coupled shear-tension loads. Constr. Build. Mater. **304**, 124517 (2021)
20. H. Do, Y.Y. Tan, N. Ramos, J. Kiendl, O. Weeger, Nonlinear isogeometric multiscale simulation for design and fabrication of functionally graded knitted textiles. Compos. B **202**, 108416 (2020)
21. J. Chen, Y. Xia, B. Zhao, W. Chen, M. Wang, F. Luo, Nonlinear characteristics and a new fractional constitutive model for warp knitted NCF composites under normal loading conditions. Polym. Test. **107**, 107464 (2022)
22. M.S. Yeoman, D. Reddy, H.C. Bowles, D. Bezuidenhout, P. Zilla, T. Franz, A constitutive model for the warp-weft coupled non-linear behavior of knitted biomedical textiles. Biomaterials **31**, 8484e8493 (2010)
23. A. Aktas, M. Aktas, F. Turan, Impact and post impact (CAI) behavior of stitched woven–knit hybrid composites. Compos. Struct. **116**, 243–253 (2014)
24. P. Justin McKee, A.C. Sokolow, J.H. Yu, L.L. Long, E.D. Wetzel, Finite element simulation of ballistic impact on single jersey knit fabric. Compos. Struct. **162**, 98–107 (2017)
25. D.-s. Li, Y. Yang, Z. Wang, L. Jiang, Experimental investigation on mechanical response and failure analysis of 3D multi-axial warp knitted hybrid composites. Compos. Struct. **246**, 112340 (2020)
26. D. Aranda-Iglesias, G. Giunta, A. Peronnet-Paquin, F. Sportelli, D. Keniray, S. Belouettar, Multiscale modelling of the mechanical response of 3D multi-axial knitted 3D spacer composites. Compos. Struct. **257**, 113139 (2021)
27. B. Pierrat, V. Novacek, S. Avril, F. Turquier, Mechanical characterization and modeling of knitted textile implants with permanent set. J. Mech. Behav. Biomed. Mater. **114**, 104210 (2021)

Knitwear Dyeing: Theory and Sustainability

Kashif Iqbal, Hafsa Jamshaid, and Rajesh Mishra

1 Introduction

Basic terminology

Preparation: "Preparation" is the term used to describe the processes which make the fabric ready for subsequent processes such as coloration and finishing. This term is also called pretreatment alternatively.

Dyeing: The process where fabric is immersed in aqueous dyeing solution/dispersion, where said dye is transferred to the textile substrate in such a way that the fabric resists the dye with certain forces to come back into the solution.

Finishing: Finishing is the process where certain mechanical and functional properties are imparted to the textile substrate to give the final touch to the fabric.

Exhaustion: The transfer of dyestuff from aqueous to the textile substrate is called exhaustion. The term "exhaustion" defines the initial strike of the dyestuff in the dyeing process.

Substantivity: The character by which the textile substrate does not leave the dyestuff, and weaker forces are responsible for this characteristic. Substantivity is a qualitative entity, unlike affinity which is expressed quantitatively. Some researchers use these two terms (substantivity and affinity) interchangeably.

Colour fastness: After dyeing, the resistance of dye to bleed against washing or rubbing is defined as washing and rubbing fastness, respectively. If the colour of the fabric does not fade in sunlight, this is attributed to the light fastness property of the dye.

K. Iqbal (✉) · H. Jamshaid
School of Engineering and Technology, National Textile University, Faisalabad, Pakistan
e-mail: kashif.iqbal@ntu.edu.pk

R. Mishra
Faculty of Engineering, Czech University of Life Sciences Prague, Prague, Czech Republic

H. Jamshaid, R. Mishra (eds.), *Knitting Science, Technology, Process and Materials*, Textile Science and Clothing Technology,
https://doi.org/10.1007/978-3-031-44927-7_6

2 Wet Processing of Knitwear

Garments made of knitted fabric are becoming popular due to their inherent properties such as casual wear, drape and elasticity, and are much demanded by the end users. Naturally, cotton contains impurities such as wax and pectin, and in addition, during knitted fabric manufacturing, needle oils are transferred to the fabric. So de-oiling process is an essential part of wet processing after which half bleaching is done for the substrate to be dyed. Full bleach is done in case white fabric is needed. Later, the dyeing process is done usually with reactive dyes in the case of cotton and dispersed dyes for polyester. All these processes are done in soft flow machines (jet dyeing machine) using the exhaust method. Different manufacturers have launched soft flow machines with the latest technology working on ultra-low liquor ratios up to 1:3 to 1:5 according to their claim. Figure 1 shows the soft flow machine used in knitwear exhaust dyeing.

Figure 2 shows the complete knitwear processing line where fabric after dyeing is finished either in tubular form or in open form as required by the customer. For open segment, fabric is first slit by a slitting machine where the sensor senses the path of the unstitched line. After that, the route for both open and tubular fabric is the same, i.e. stretch dry compact as shown in Fig. 2. The purpose of stretching process is to overcome the lengthwise negative shrinkage taking place during fabric manufacturing and dyeing process. Hence, overfeed is given lengthwise while stretching the fabric widthwise to control the shrinkage/dimensional stability. During stretching process, the application of chemicals takes place in the padder such as softener, antibacterial finish, or resin according to the requirement of the customer.

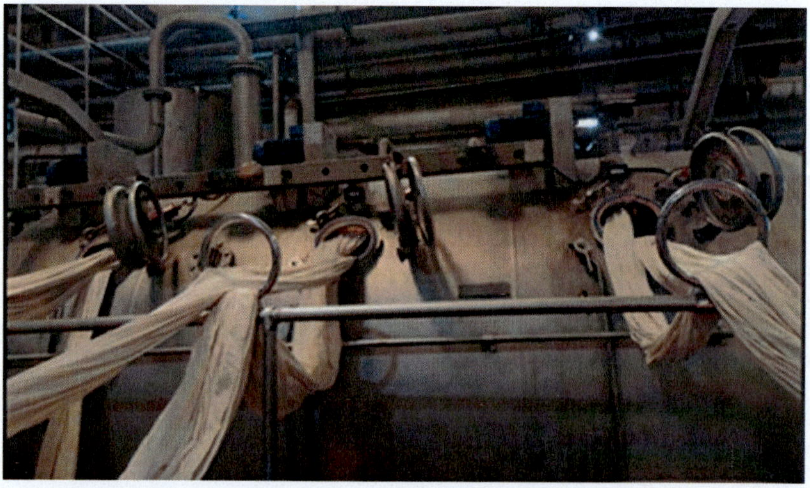

Fig. 1 Soft flow jet dyeing machine

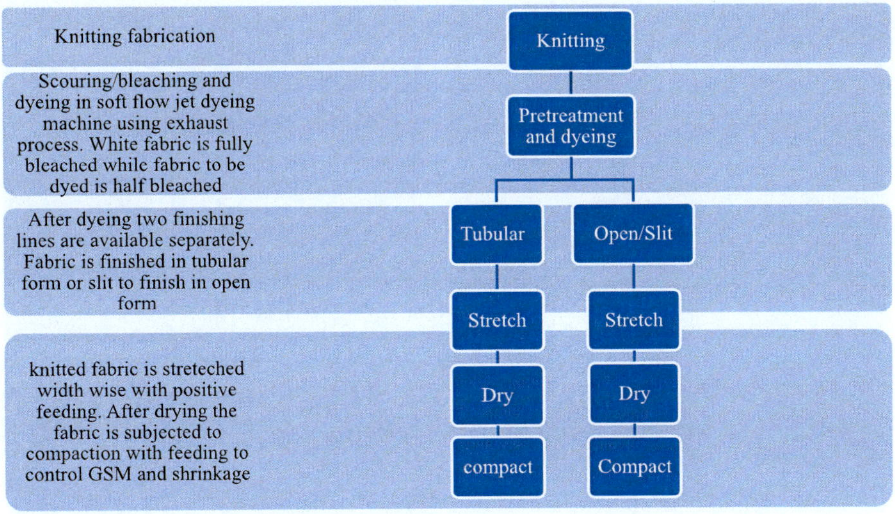

Fig. 2 Knitwear wet processing flow chart

Fig. 3 Finishing of tubular knitted fabric

Figure 3 shows the diagram of finishing process of knitted fabric in tubular form. The fabric as mentioned earlier is stretched widthwise in tubular form after the application of chemical in the padder and accumulated in a trolley followed by drying in an oven. The last process of compaction takes place where fabric is compacted lengthwise to control shrinkage and weight (gram per square meter (GSM)) of the fabric. During compaction, steam is provided to moisten the fabric for better dimensional stability.

Figure 4 shows the same finishing process of open knitted fabric where tubular fabric is slit to bring it into open form. Then, the fabric is stretched after padder application (usually softener or water) with overfeed as shown in Fig. 4. Because the fabric is stretched a lot widthwise, overfeed is required to avoid any damage or

Fig. 4 Finishing of open knitted fabric

tearing of fabric. After stretching, the fabric is dried and then compacted. The process of compaction is the same as in tubular form except that a Mahlo system is attached to the compact machine (sometimes also called Sanforizing machine). The Mahlo system is used for the detection and control of torque, bow and skewness of the fabric.

3 Theory of Dyeing

The focus of this chapter is to provide theoretical knowledge of dyeing along with its importance and application in practical dyeing. Theory of dyeing is elaborated keeping in mind the exhaust dyeing system for knitwear using anionic dyes (direct, reactive, vat and sulphur) for cellulosic. However, this theory can form the basis for other types of dyeing systems but that needs a separate chapter for discussion and will be covered in the upcoming chapter.

3.1 *Phases of Dyeing*

There are five phases in the dyeing process as shown in Fig. 5. The dyer needs to keep in mind all these phases when deciding parameters or defining the route of dyeing for any textile material [7].

3.1.1 Disaggregation of Dyestuff

Dye molecule in a solution does not exist in mono-molecular form, and because of certain weaker forces found among dye molecules, they tend to aggregate, which change the thermodynamical behaviour of dyestuff. The dyestuff in aggregated form has less solubility and hence tends to rush towards fibre increasing the initial strike. This effect leads to unlevel dyeing and even becomes more severe if the concentration of dyestuff is higher and liquor ratio is low. Hence, for a dyer to acquire level dyeing, the most important factor is disaggregation of dyestuff in solution/dispersion form by taking necessary steps which usually involves careful dye liquor preparation from paste with mechanical agitation as well as adding auxiliaries which can reduce the aggregation of dyestuff and enhance solubility.

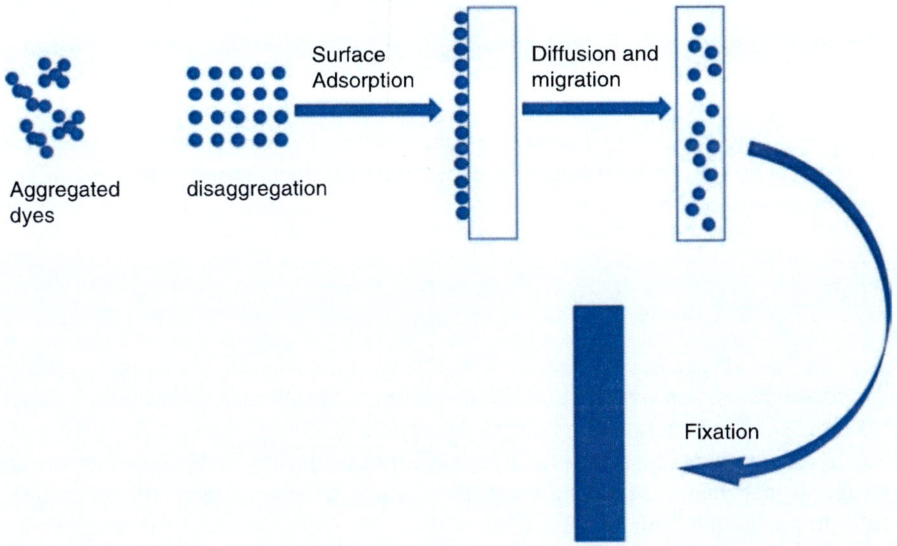

Fig. 5 Phases of dyeing

3.1.2 Exhaustion

In exhaust dyeing, exhaustion refers to the transfer of dyestuff molecules from the solution to the fabric. Exhaustion is a function of time and increases with the increase in concentration of dyestuff and electrolyte. On the contrary, exhaustion decreases with an increase in solubility and temperature. Therefore, exhaustion can be controlled with above-mentioned parameters and if required, temperature can help to improve the initial strike for level dyeing. The dependence of exhaustion on solubility and aggregation is explained in Sect. 3.2.

3.1.3 Adsorption

Rate of adsorption depends upon the rate of exhaustion. This third phase of dyeing is very important in practical dyeing. Hence, the route of dyeing is designed in such a way that maximum rate of adsorption is achieved. Two important factors are involved in adsorption of dyeing: ramp (temperature gradient) and the addition of salt to get maximum exhaustion on the fabric. The substantivity is also an important factor which helps the dyestuff stay on the fabric.

3.1.4 Diffusion

The rate of diffusion or penetration is a key factor to determine the rate of dyeing. Once the adsorption is achieved, the next phase of diffusion occurs where optimum temperature is provided for penetration of dyestuff within the fibrous polymer. Rate of diffusion increases with increase in temperature and time. Time is a vital factor for efficient migration (distribution of dyestuff within the fibre in exhaust dyeing) of dyestuff to get level dyeing.

3.1.5 Fixation

Here the final stage comes where fixation of dyestuff occurs. In case of reactive dyeing, fixation is done providing alkaline condition to the fabric. The addition of alkali is very important to get maximum dyestuff fixed to the fabric rather than hydrolysis. The hydrolysis not only wastes the dyestuff but also attaches itself to the fabric with weaker forces, hence, requiring extensive soaping to remove them for better fastness properties. Nowadays, high-fixation dyes are available in the market, and the manufacturer of such dyes claims their fixation level is above 90%. Hence, all the above phases are connected and are critical to be considered by dyers in the same order as mentioned for level and sustainable dyeing.

3.2 Solubility and aggregation

Aggregation of dyestuff has an important factor in dyeing which depends upon the molecular structure complexity and solubility. If the dye has good solubility (attachment of enough solubilizing groups), its aggregation will be less resulting in controlled exhaustion. The graph of exhaustion is given in Fig. 6 illustrating as a function of time.

The graph in Fig. 6 shows that exhaustion increases as a function of time. Maximum exhaustion is achieved within first 20 min which indicates that initial strike of the dyestuff is usually very high. Hence, if dye is more in concentration, the aggregation occurs which assists the dye to exhaust quickly further raising its strike. After a certain time, an equilibrium is attained, and there is insignificant change in exhaustion.

Salt (common salt or Glauber's salt) is used as an exhausting agent in exhaust dyeing of cotton. It is recommended to use salt after a prescribed time once the natural exhaustion decreases, so that the initial strike of dyes can be controlled for even dyeing shade.

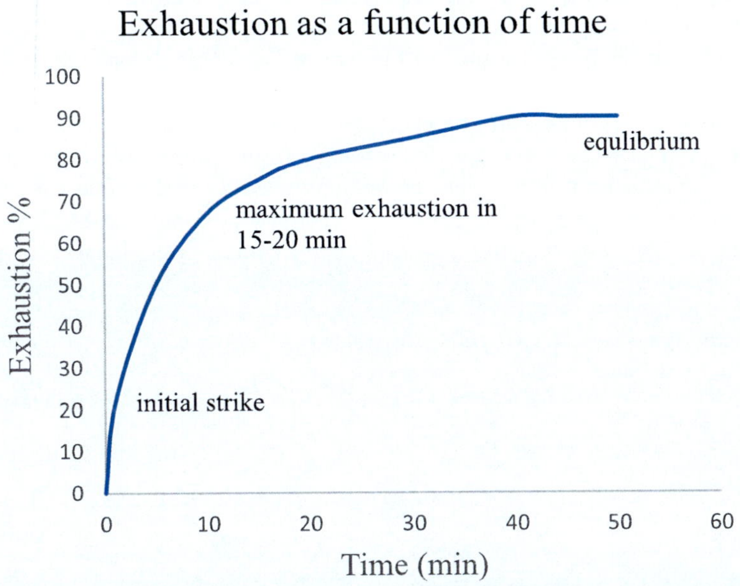

Fig. 6 Exhaustion of dyestuff from liquor to the textile substrate

3.3 Chemistry of Dye–Fibre Interaction

The relation between dye and fibre depends upon the physico–chemical interaction between dyes and fibre which further depends on the chemistry of dyes. The dye manufacturer always bears in mind the chemistry of dyes in relation to the fibres while synthesizing them. Hence, dyestuff manufacturing is a crucial phenomenon where substantivity, solubility, aggregation and levelling are associated with the chemistry of dyestuff. Therefore, dyers need to have some knowledge about the chemistry of dyes so that they can design the dyeing route and establish parameters after looking at the chemical structure of dyes. The relation is explained here by showing (Fig. 7) two different structures of anionic dyes for cellulosic.

In Fig. 7, two direct dyes are shown having different number of solubilizing groups. Color index (CI) direct Blue 1 will impart more solubility due to six ionic groups (sulphonic and amine groups) which lessen the dye aggregation and keep the dyestuff in mono-molecular form. While CI direct Red 28 also has enough ionic group content to impart good solubility, the opposite charge located at the para position could stack the dyestuff into a macromolecular structure leading to high aggregation of the dyestuff. This aggregation of dyestuff not only impairs the level dyeing due to high initial strike but also decreases the diffusion of dyes in the fibre due to its macro structure.

To further illustrate the dye–fibre relationship, substantivity is worth discussing at this point. Substantivity as described earlier is a character by which the dyestuff with certain forces stays on the surface of fibre and does not leave it. These forces are usually weaker forces in most of the dyes except ionic dyeing where opposite charges are responsible for attraction (e.g. dyeing of nylon with acid dyes or acrylic with cationic dyes, etc.), and this character is referred to as affinity. The weaker forces are van der Waals forces which originate from different sources such as dispersion and induction effect. The major binding force responsible for dye–fibre interaction is hydrogen bonding. The substantivity is controlled by the dyestuff structure and depends on certain factors such as linearity/coplanarity of the dye

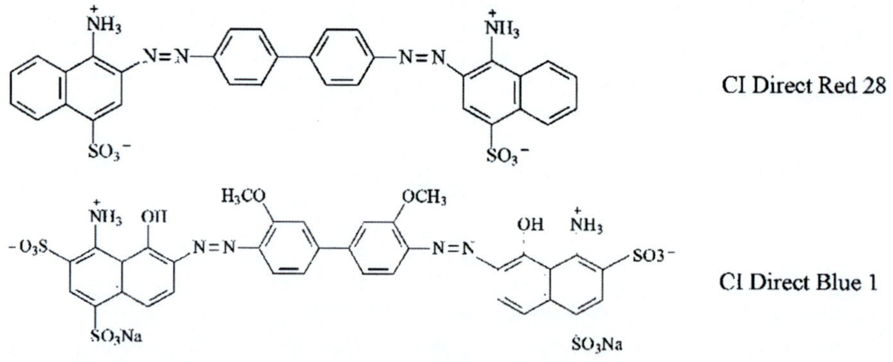

Fig. 7 Dyestuff chemistry affecting dyeing phenomenon

structure, groups capable of forming hydrogen bonds and minimum possible number of solubilizing groups.

It is well known that most of the fibrous polymers have linear configuration which require the dyestuff structure to be linear to establish effective bonding. If dyestuff structure is non-coplanar and non-linear, its attachment with the fibrous polymer with weaker forces will be to a lesser extent. Similarly, fibres such as cellulosic have –OH groups which can form hydrogen bonds with the dye structure. Hence, the dye structure should contain maximum groups which are capable of forming hydrogen bonds with fibres. The third parameter that helps to confer substantivity is the presence of minimum solubilizing groups so that once the dyestuff is adsorbed on the fibre surface, the dyestuff should not diffuse back into the solution [25].

3.4 *Kinetics of Dyeing*

From the above discussion, the reader can understand that dyeing is a complex phenomenon and requires comprehensive knowledge of theory for practical dyeing. For example, increase in solubility causes decrease in exhaustion and substantivity which can affect the rate of adsorption resulting in the reduction of rate of dyeing. The higher substantive dyes can lead to more aggregation which can affect the migration of dyes within the fibre and rate of penetration of the same. An addition of salt could help to overcome the repulsive forces between negatively charged dyes and fibres and can also enhance dye aggregation. Similarly, increase in temperature can increase the rate of diffusion and dyeing and can also increase the rate of hydrolysis. So, a dyer should design the parameters and route of dyeing keeping dyeing theory and kinetics in mind. A detailed discussion of physical chemistry and kinetics of dyeing is out of scope of this chapter; however, a brief discussion is given for consideration.

The rate of dyeing is determined by the rate of diffusion of dye within the fibre which is affected by different factors that tend to decrease its rate. Among all factors, worth mentioning ones are state of aggregation of dyes which prevents the diffusion of dye as they are in a macromolecular state and hence stay on the surface of the fibrous material which could diminish the rate to zero value. The fibrous amorphous part and electrical repulsion between dye and fibre are also decisive and influential in controlling the kinetics of dyeing. The dyer needs to consider such factors and prevent the system from having state of aggregation and electrical repulsion because molecular structure is a major cause of aggregation as already shown in Fig. 7, and the same is illustrated in Fig. 8 where CI Direct Red 23 has 400 times greater time of half dyeing than stilbene-based (CI Direct yellow 12) dye offering a low rate of diffusion.

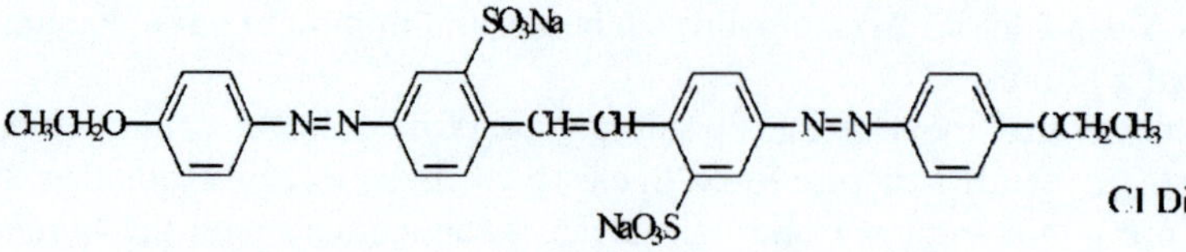

CI Direct Yellow 12

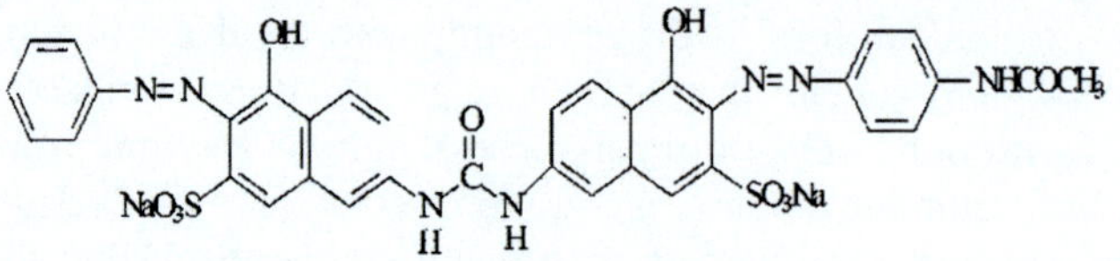

CI Direct Red 23

Fig. 8 kinetics of dyeing with their structures

Furthermore, the kinetics of dyeing are also influenced by:

- Concentration of dyes.
- Affinity of dyes.
- Electrolyte concentration.
- Temperature of dyeing.
- Physical structure of fibres.

To understand the influence of concentration of dyes, it is suggested to read isotherms because different dyes for various fibres show changed behaviour, e.g. the increase in concentration of disperse dyes for polyester does not have significant effect on diffusion, whereas direct or reactive dyes for cotton show linear effect while ionic dyes for nylon and acrylic increases the rate parabolically. The affinity of dyes has inverse relation with the rate of diffusion; as affinity of dyes increases, the resistance of dyes to diffuse in the fibre also increases, hence, the rate of diffusion decreases.

The concentration of salt increases the rate of diffusion of dyes for cellulosic as the electric repulsion decreases between dye and fibre due to addition of salt. However, after an optimum level, the rate of diffusion decreases sharply because salt increases the aggregation and affinity of dyes. Similarly, increase in temperature increases the rate of diffusion because of increased kinetic energy of the dye molecules even if they are in the macromolecular state. The physical state of the fibres also influences the rate of diffusion as dyeing occurs at amorphous regions, hence, the ratio of amorphosity to the crystalline region is important for dyeing kinetic. Therefore, the cotton is mercerized to increase the dyeability of fibre, and synthetic fibres are subjected to thermal treatment where the chains are rearranged and mobile in the amorphous region due to the modification in glass transition temperature resulting in the increase of diffusion rate [12].

4 Dyeing Theory to Practical Application

Dyeing theory has its real importance in practical dyeing, but it has been exploited only in two manufacturing industry called dyestuff manufacturing and machine manufacturing. This is the reason why technological developments in dyestuff and machine manufacturing succeeded in solving current issues such as high fixation problem in dyeing, high shade depths, dyeing with better characteristics (fastness) and low consumption of water and energy in the machine. For example; the problem of high liquor ratio resulted in the advancement of jet machine with low liquor ratios. Nowadays, even soft jet machines are available which are claimed to perform under 1:5 goods–to-liquor ratio. Unfortunately, the dyers do not utilize the theory in true sense due to which hit and trial methods are still in use to solve the problem in practical dyeing (James [20]). However, in industry, successful implementation of theory to practical dyeing is exploited to a little, which includes the:

- Control on levelling by controlling the parameters such as temperature, pH, premeditated amount of auxiliaries and metered chemical addition.
- Selection of dyes in such a combination that has compatibility in terms of substantivity, exhaustion, levelling and fixation to avoid any problem in practical dyeing. Different dyestuff manufacturers have designed computer-aided procedures using spectrophotometers which take into account the compatibility of dyes to be used. In 1990, Badische Anilin- und Sodafabrik (BASF) defined the reactive dye compatibility matrix (RCM) encompassing five key parameters such as substantivity, exhaustion, fixation, migration index and levelling factor. Such machines are in market nowadays which are automated with computer-aided software, showing that real-time exhaustion and kinetics can be controlled during the process which can prevent the faulty dyeing.
- Proper selection of machine parameters to have appropriate liquor exchange such as flow rate and circulation time.

The above-mentioned theory will help the dyer to consider important factors while designing their routes for exhaust dyeing which will help them achieve right first-time approach in dyeing and laboratory–to-bulk reproducibility.

5 Sustainable Approaches in Dyehouse

Sustainability is a major topic of concern nowadays and deals with all kinds of problems related to environmental, economic and social concerns. Among all manufacturing industries in the world, textile sector specially dyehouse is considered to be the most polluting industry [4]. A typical wet processing industry produces effluent containing high biological oxygen demand (BOD), chemical oxygen demand (COD), total dissolved solids (TDS), pH and colour. Moreover, the immense use of water in textile wet processing industry creates scarcity of water which not only has social impact but will also affect the industry in near future leading to economic issues [24]. This chapter deals in knitwear dyeing, and the most widely used process in knitwear dyeing is the exhaust dyeing process, where in addition to dyes and auxiliaries, plenty of salt is used, generating high TDS value in effluent. Furthermore, commonly used dyestuff in exhaust cotton dyeing is reactive dye due to its brighter colour, fastness properties and ease in application, which also gets hydrolysed leaving an unfixed colour to the effluent. Hence, the appropriate use of dyes and chemicals along with the right dyeing profile is needed to avoid waste and rigorous effluent [27]. There are several strategies that can be applied, but here, only three approaches are described, which, if adopted, may lead to sustainability in dyehouse, such as:

Right first-time approach (RFT).
Material and process modification.
Water conservation.

5.1 *Right First-Time Approach*

Right first-time approach defines the accomplishment of dyeing process without the need of reprocessing such as correction of shade and addition of dye. The main objectives for a dyer are to achieve dyeing of a material with correct target colour containing good fastness properties, minimum cost of dyeing and production in time without delay. These all can be achieved using right first-time approach rather than blind dyeing.

5.1.1 Impact of Dyeing Theory

Dyeing theory has already been explained in the previous section and is essential to utilize during the dyeing procedure along with spectrophotometer if possible. The achievement of right first-time is only possible if the selection of liquor ratio, time, temperature and auxiliaries is in concordance with exhaustion, levelling, substantivity, diffusion and fixation of dyestuff [3]. Hence, the progress of dyeing is based on optimization of the dyeing process with appropriate selection of parameters keeping physico–chemical aspects of dyes, auxiliaries and substrate in consideration [6, 13].

5.1.2 Impact of R&D

Research and development (R&D) always have an impact on the progress of any process. It includes the selection of dyes and chemicals with proper knowledge of dyeing profile using design of experiments. These repeated trials help to get accuracy with reproducibility at laboratory scale and its transfer to bulk with laboratory-to-bulk reproducibility. Furthermore, repeated trials in bulk can promote reproducibility within the same colour (James [19]). Here, it is worth mentioning the use of computer colour matching (CCM) for the prediction of dyes and recipes as well as for the implementation of a better colour quality control system [14]. Park and Shore ([21]) mentioned important factors in their article as mentioned below, which can influence RFT dyeing and should be monitored either by laboratory check or by establishing standard operating procedures.

- Materials such as water, substrate, dyes and chemicals should be pure and well prepared with predetermined moisture content.
- For the dyeing process, weight of substrate to be dyed and weight of dyes and chemicals should be accurately measured with proper dispensing system. Control of liquor ratio, pH, time, temperature and liquor or substrate flow should be figured out using standard operating procedures.
- For colour control, a proper colour quality control system should be implemented which includes the selection of dyes in combination, their profile compatibility, accurate transfer of dyeing recipe from laboratory–to-bulk reproducibility. Correct SOPs should be prepared and implemented with substantial training of laboratory and bulk-dyeing operators.

5.1.3 Implementation of RFT Using Six Sigma

Prashar [23] performed a case study in a dyeing mill to investigate the factors responsible for harming the RFT dyeing process and identified the root causes by adopting six sigma: define, measure, analyse, improve and control (DMAIC), technique. In his investigation, he found "shade mismatch" was the major parameter which hindered achieving the objective of RFT. The analysis further revealed that fabric dyeability, dyestuff strength variation and water quality were the main root causes for the problem. The study improved the RFT dyeing up to 4% with a cost saving of Indian rupees 2.95 million per month. He concluded that six sigma methodology is a very useful approach which, if implemented, can minimize the process variation and defects which usually originate due to complex dyeing phenomenon.

5.2 *Material and Process Modification*

5.2.1 Process Modification

In this section, material and process modification will be dealt for the establishment of eco-friendly dyeing system. In cotton reactive dyeing, plenty of salt is used which later becomes a part of effluent, resulting in high value of TDS. Surface modification of cotton is performed using different chemicals such as 3-chloro-2hydroxypropyl trimethyl ammonium chloride (CHPTAC). After treatment with this chemical, cotton becomes cationized and develops affinity for anionic dye and does not need the addition of salt [1]. Similarly, plasma treatment can modify the surface chemistry of cotton, and it can enhance the dyeing capability with better fastness characteristics. Correia et al. used different cationizing agent and plasma treatment containing helium and oxygen gas on cotton fabric for surface treatment. This treatment eliminated the pretreatment process such as scouring and bleaching along with high adsorption of anionic (reactive and acid) dye without the addition of salt. The dyed materials showed better fastness properties with high K/S (absorption coefficient (K) and scattering coefficient (S)) value as well as level dyeing was achieved [9]. It was reported that hydrophilicity caused by plasma treatment on cotton fabric improves the wettability and adsorption of dyes. Further, it was investigated that exhaustion of reactive dyes increased up to 10% for cotton dyeing and 15% increase was in K/S value [22].

Cold pad batch is a method where pretreatment and dyeing are performed using pad technique, and batch is stored for some time with a polyethylene covering to provide reaction time. The extra need of water and energy required to dry the fabric is eliminated, and it is claimed that 50% of water can be saved in comparison to the conventional methods [8]. As already mentioned, reactive dyes are very famous for cotton due to their fastness properties and ease in application. Unfortunately, reactive dyes have one disadvantage of being hydrolysed in water owing to which their exhaustion decreases. Different methods were developed to overcome this problem, and finally cold pad batch method was developed for reactive dyes where not only the chances of hydrolysis decrease but also the use of salt eliminates completely [18]. The cold pad method is usually famous for woven fabric; however, knitwear pretreatment and dyeing have also started in few countries recently and under development. This method is highly sustainable as the use of water and energy is reduced enormously, but it also has a disadvantage of having a large time for batching [5].

5.2.2 Dyes and Auxiliaries

The use of eco-friendly salt instead of common salt or Glauber's salt has also been reported by different researchers such as magnesium citrate, magnesium acetate, magnesium polyacrylate and trisodium citrate. These salts are biodegradable unlike the conventional salt and can be removed from effluent using precipitation method by adjusting the pH of the solution [10, 15].

Natural dyes are as old as the human civilization and only one century ago, the synthetic dyes overtook the dyes from natural resources. Natural dyes are claimed to be non-hazardous, having little impact on environment and waste disposal. Natural dyes can be extracted from different natural resources such as plant leaves, stem, root, flowers, minerals and insect secretions. Chemically, they are quinones, benzoquinones, naphthoquinones, anthraquinones, flavanol, flavanones, indogoid, chlorophyll etc.; however, whether natural dyes can replace the synthetic dyes in terms of their colour gamut, application ease, fastness properties and sustainability [2] or not is still a question. Fashion industry is demanding the use of natural dyes instead of non-eco-friendly synthetic dyes to claim sustainability. Dyeing with dyes from different natural resources still needs an extensive R&D to optimize the dyeing conditions such as requirement of pH, temperature, liquor ratio, etc. as these dyes are suitable for certain fibres while in some cases, they need strong mordant to achieve application characteristics [26].

The development in natural dyeing is a hot topic today, and many researchers are working on it. Dyes are extracted from different natural resources to complete the colour gamut, and furthermore, their optimization in a dyeing process such as to achieve high exhaustion (specially in darker shade), high fixation and good fastness properties are still in progress. Recently, Nawaz et al. [16] dyed wool fibre with natural dye extracted from *Dalbergia sissoo* using different metal salts as mordants. They found that high K/S value can be achieved with good fastness properties containing ferrous sulphate as a mordant using meta mordanting technique. The use of natural mordants instead of metal salt is also reported in literature where researchers used *Aloe vera*, lemon and *Embiclica officinalis* G, etc. as biomordant in dyeing with different natural dyes. Nilani et al. [17] extracted dye from Marigold flower to dye different synthetic fibres including human hair using aloe vera juice as mordant in comparison with other metallic mordants and found promising results.

5.3 Water Conservation

Water is a necessity of textile industry especially wet processing where large amount of water is used. Due to growing fashion and living standards, the increase in production and fashion garments is increasing day by day. Almost 150 litre of water is used per kg of the fabric in whole supply chain which is attracting the attention of researchers and manufacturers to put their resources and efforts on water conservation [11]. This has led to the advancement in technology in all aspects such as the development of dyes and auxiliaries, development in processes and improvements in machine technology. Former techniques have been discussed to some extent in previous sections while machine development for knitwear dyeing is briefly described here.

Initially, winch machines were used for dyeing of knitwear fabric which use liquor from 20 litres to 40 litres per kg of the fabric. The development in jet dyeing machine reduced the requirement of water and in early stages; up to 10 litres of liquor was used per kg of the fabric. Later, different machine manufacturing companies developed the technology of low liquor machines called soft flow and air flow dyeing machines. Fong's, one of the largest manufacturers of soft flow machines, developed this technology by circulating the liquor due to which penetration of liquor increases within the fibrous polymer system. Later, the modification to air flow machine was done which requires the air to flow the fabric, reducing the need for extra liquor. They claim their dyeing machine works on ultra-low liquor ratios up to 1:2 or 1:3 which not only minimizes the water consumption but also increases energy saving up to 40% [11].

In textile dyeing, large amount of water is used during washing specially in reactive dyeing where unfixed dyes become substantive and attach themselves to the fabric with weaker forces. Plenty of water is used to remove the unfixed dyes after dyeing which not only increases the consumption of water but also the cost of washing. The following can reduce the consumption of water resulting in low cost of washing.

- Use of high exhaustion and high fixation dyes.
- Continuous washing for knitwear fabric instead of washing in exhaust machine.
- Counter-flow washing techniques where only one washer is filled with fresh water while water is transferred to other boxes due to gravitational forces. Furthermore, efficiency of washing in counter-flow also increases due to more agitation and turbulence. Even a factory using exhaust machinery for knitwear dyeing can adopt the continuous counter-current washing system for knitted fabric.
- Reuse of suitable dyebath.
- Use of water softening plant and its scheduled maintenance.
- Wastewater treatment and reuse of treated water.
- Operators training on optimum use of water and monitoring of water lines with installed meters.

6 Conclusion

Dyeing is a complex phenomenon, and dyers are required to have basic knowledge of dyeing theory and kinetics by using optimum parameters and designing appropriate routes to avoid faulty dyeing. To achieve laboratory–to-bulk reproducibility, the optimization of the levels of dyeing parameters is essential, and R&D is required in industry at laboratory and bulk scale. If dyeing theory is accurately practiced, the repeatability in the dyeing result such as shade depth, shade match and fastness properties is possible in bulk. For designing the dyeing route and profile, phases of dyeing such as disaggregation, exhaustion, adsorption, diffusion and fixation should be well understood. Furthermore, to compete in the global market, an effort to attain sustainability and an investment on R&D is inevitable which will surely enhance the process quality and productivity.

References

1. M.S. Atiq, A. Rehman, K. Iqbal, F. Safdar, A. Basit, M. Ashraf, et al., Salt free sulphur black dyeing of cotton fabric after cationization. Cellul. Chem. Technol. **53**(1–2), 155–161 (2019)
2. T. Bechtold, R. Mussak, *Handbook of Natural Colorants*, vol 8 (Wiley, Hoboken, 2009)
3. C. Bird, P. Rhyner, The dyeing of cellulose acetate with disperse dyes X—Saturation values with mixtures of dyes. J. Soc. Dye. Colour. **77**(1), 12–16 (1961)

4. R. Blackburn, *Sustainable Textiles: Life Cycle and Environmental Impact* (Elsevier, London, 2009)
5. A.D. Broadbent, X. Kong, The application of reactive dyes to cotton by a wet-on-wet cold pad-batch method. J. Soc. Dye. Colour. **111**(6), 187–190 (1995)
6. B.C. Burdett, A.J. King, The dyehouse into the 21st century. Energy **9**, 14 (1999)
7. J. Cegarra, P. Puente, J. Valldeperas, *The Dyeing of Textile Materials: The Scientific Bases and the Techniques of Application* (Textilia, Biella, 1992)
8. L. Chen, Cold pad-batch technique of energy conservation and emission reduction (1). Text Finish J **31**(11), 51–54 (2009)
9. J. Correia, K. Mathur, M. Bourham, F.R. Oliveira, R.D.C. Siqueira Curto Valle, J.A.B. Valle, A.-F.M. Seyam, Surface functionalization of greige cotton knitted fabric through plasma and cationization for dyeing with reactive and acid dyes. Cellulose **28**(15), 9971–9990 (2021)
10. S.A. Farha, A. Gamal, H. Sallam, G. Mahmoud, L. Ismail, Sodium edate and sodium citrate as an exhausting and fixing agents for dyeing cotton fabric with reactive dyes and reuse of dyeing effluent. J. Am. Sci. **6**(10), 109–127 (2010)
11. M. Gopalakrishnan, V. Punitha, D. Saravanan, Water conservation in textile wet processing, in *Water in Textiles and Fashion*, (Elsevier, London, 2019), pp. 135–153
12. A. Johnson, *The Theory of Coloration of Textiles* (Society of Dyers and Colourists, Bradford, 1989)
13. J. Keaton, B. Glover, A philosophy for dyeing in the next decade. J. Soc. Dye. Colour. **101**(3), 86–98 (1985)
14. M.R. Luo, Development of colour-difference formulae. Rev. Prog. Color. Relat. Top. **32**, 28–39 (2002)
15. S.B. Moore, Low toxicity, biodegradable salt substitute for dyeing textiles: Magnesium acetate in direct or reactive dyeing of cotton. Google Patents (1993)
16. N. Nawaz, A. Rehman, M.T. Hussain, F. Safdar, K. Iqbal, Dyeing of wool with *Dalbergia sisso* as an eco-friendly substituent of conventional hazardous synthetic dye. J. Nat. Fibers **19**(15), 10068–10081 (2022)
17. P. Nilani, B. Duraiswamy, P. Dhamodaran, N. Kasthuribhai, S. Alok, B. Suresh, A study on the effect of marigold flower dye with natural mordant on selected fibers. J. Pharm. Res. **1**(2) (2008)
18. W. Nitayaphat, P. Morakotjinda, Cold pad-batch dyeing method for cotton fabric dyeing with Uncaria gambir bark using ultrasonic energy. Chiang Mai J. Sci. **44**, 1562–1569 (2017)
19. J. Park, Dyeing laboratory developments. Rev. Prog. Color. **34**, 86–94 (2004)
20. J. Park, J. Shore, Does dyeing practice need dyeing theory? Color. Technol. **123**(6), 339–343 (2007)
21. J. Park, J. Shore, Evolution of right-first-time dyeing production. Color. Technol. **125**(3), 133–140 (2009)
22. A. Patino, C. Canal, C. Rodríguez, G. Caballero, A. Navarro, J.M. Canal, Surface and bulk cotton fibre modifications: Plasma and cationization. Influence on dyeing with reactive dye. Cellulose **18**, 1073–1083 (2011)
23. A. Prashar, Right-first-time dyeing in textile using Six Sigma methods. Int. J. Sci. Eng. Res. **4**(8), 1517–1525 (2013)
24. R. Ramasany, H.A.M. Ahmed, S.S. Karthik, Development of microbial consortium for the biodegradation and biodecolorization of textile effluents. J. Urban Environ. Eng. **6**(1), 36–41 (2012)
25. J. Shore, *Cellulosics Dyeing* (Society of Dyers and Colourists, Bradford, 1995)
26. P.S. Vankar, Chemistry of natural dyes. Resonance **5**(10), 73–80 (2000)
27. G. Varadarajan, P. Venkatachalam, Sustainable textile dyeing processes. Environ. Chem. Lett. **14**, 113–122 (2016)

Textile Testing and Quality Control in Knitting

Hafsa Jamshaid and Rajesh Mishra

1 Introduction

Weft-knitted fabric is usually knitted on a double jersey circular machine, and the fabric width is limited by the needle cylinder [1–3]. The two primary forms of weft knitting machines are circular knitting machine and flat knitting machine. Circular machines can be subdivided into single jersey, dial and cylinder, and double cylinder purl machines according to the needle set used and the fabrics made [4]. Weft knitting machines with two sets of needles have the potential to produce two separate covering layers that are held together by tucks. It is considered that dial and cylinder, and purl machines are able to produce knitted fabric. Flat knitting machines can be divided into two types – the V-bed machine and flat purl machine. V-bed machine is useful in the manufacture of knitted fabrics, while flat purl machine is rarely used in today's applications. The dial and cylinder machine can connect two separate layers of fabric together by the use of various combinations of stitches. To produce a dial and cylinder–knitted fabric, at least three different yarns are required to form each course of the fabric including yarn for dial needles, yarn for cylinder needles, and knitted yarn [5, 6]. Dial height determines the amount of pile yarn being fed between two surface layers. By adjusting the dial height, the producer can alter the distance between the two layers [7, 8]. Knitted fabrics can also be produced with a V-bed flat knitting machine. A tubular knitted fabric is connected with mainly monofilament pile connections. Pile yarns are inserted with a zigzag movement between two fabric layers. The angle of connections can be varied, which enables a

H. Jamshaid
School of Engineering and Technology, National Textile University, Faisalabad, Pakistan

R. Mishra (✉)
Faculty of Engineering, Czech University of Life Sciences Prague, Prague, Czech Republic
e-mail: mishrar@tf.czu.cz

H. Jamshaid, R. Mishra (eds.), *Knitting Science, Technology, Process and Materials*, Textile Science and Clothing Technology,
https://doi.org/10.1007/978-3-031-44927-7_7

construction with a localized adjustment of compression stiffness. The distance between the two needle beds determines the knitted fabric thickness. Unlike circular knitting machines, the distance between the two needle beds of a flat knitting machine is fixed around 4 mm. By using a computerized flat knitting machine with elastomeric yarn, the knitted fabric thickness can vary in a wide range. However, the productivity is very low while knitting the thicker knitted fabrics. The mechanism of tucking on two sets of needles leads to ineffective constraints on knitted yarn provided by outer fabric layer stitches. So, the distance between two needle beds is the cause of limited dimensions [9]. By the use of a V-bed machine, two independent covering layers are knitted on the front and back needle beds. Knitted fabric is created by tucking a pile of yarn to link the two separate fabrics together [10, 11].

Warp-knitted fabric is usually knitted on a rib raschel machine with two needle bars and a number of guide bars. The warp-knitted fabric has a higher thickness [12]. There are two major classes of warp knitting machines: raschel and tricot. Fabric on raschel machines is drawn downward from the needles almost parallel to the needle bar, at an angle of 120–160 degrees. This angle creates a high take-up tension, particularly suitable for open fabric structures such as laces and nets. The warp beams are arranged above the needle bar and centered over the top of the machine so that the warp yarns pass down to the guide bars on either side of them. The guide bars are numbered from the front of the machine. Raschel machines can accommodate at least four 32-inch diameter beams or a large number of small-diameter beams. Raschel machines typically knit with latch needles or compound needles. Machine gauge is expressed as needles per inch. The gauge range can be from 1 to 32. The simple knitting action and the strong and efficient take-down tension make the raschel machine well suited for the production of coarse gauge openwork structures using pillar stitch, inlay lapping variations, and partly threaded guide bars. Raschel sinkers perform the function of holding down the loops while the needles rise [13].

The customer requirement for the abovementioned applications depends on various factors, mainly shearing, excellent cushioning, and conditioned air and heat (breathability). There are a number of materials and structures with the abovementioned features for those applications. Airbags, bubble films, rubberized fiber cushioning, and polymer-based foams are just a few typical examples. The use of foamed materials results in a significant improvement in the passive safety, owing to their excellent energy dissipation properties. In addition, they have low apparent density, are relatively cheap, and allow great design flexibility as they can be easily modeled in complex geometric parts. However, despite their promising applications, these materials are not suitable for many critical applications due to inferior comfort properties and environmental hazards in terms of both production and recycling. Hence, in order to overcome all these drawbacks in car interior application, three-dimensional (3D) knitted fabrics has attracted the attention of researchers in recent times. Knitted fabrics are a class of material with unique properties and applications. They have two outer surfaces connected to each other with knitted yarns; they provide lightweight and bulkier structure. They are lightweight and designed to undergo very large deformations. Their properties are the results of the knitted fabric microstructure, a complex three-dimensional network, low density, and possibly high thickness, which undergo larger deformations during mechanical loading. Their compression and comfort characteristics are also better than conventional textile structures.

2 Properties of Knitted Fabrics

2.1 Mechanical Properties

The critical mechanical properties of knitted fabrics are those related to tensile strength, tear strength, and stiffness. Tensile strength of knitted fabrics measures the fabric's ability to resist the tensile forces resulting from prestress in combination with external loads, and it measures the level of direct pull force required to rupture the fiber of the material [14, 15]. Stiffness is, of course, related to modulus of elasticity of the material and the area of fibers employed, which may vary in the warp and fill directions of the material. In addition, both the type of weave employed and the manufacturing process effect stiffness variation under load due to crimp interchange.

2.2 Impact Properties

The structural parameters of a knitted fabric have a significant effect on its protective performance [16]. Among a group of knitted fabrics, the fabric knitted with higher inclination and coarser knitted monofilaments, a bigger fabric thickness, and a more stable outer layer structure has a better force attenuation capacity, Researchers have studied the impact properties of warp-knitted fabric by varying different parameters. First, the thickness and outer layer stitch density of the two fabrics are kept nearly the same. The knitted fabric with the coarser knitted monofilament has a lower peak transmitted force and a longer time to the peak point and, therefore, has a better impact force attenuation property [17, 18]. They also investigated a group of three fabrics with the same outer layer structure (chain plus inlay), and the same knitted monofilament yarn but with different knitted monofilament inclinations (underlapping one needle, two needles, and three needles between the front- and back-needle bars) is used to analyze the effect of the knitted inclination on the impact force attenuation properties of warp-knitted fabrics. The fabric thickness and stitch density of the outer layers are kept nearly the same. The number of needles underlapped determines the knitted monofilament inclination and length. The higher the number of the needles underlapped, the longer and more inclined the knitted monofilaments.

The transmitted force–time curves of these fabrics in a single layer under impact at a kinetic energy of 5 Joules are used as an example for discussing the effect of the knitted monofilament inclination with the same impact energy. It can be seen that the duration from the beginning point, where the striker contacts the fabric's upper surface, to the peak point, where the transmitted force reaches the maximal value, increases as the knitted monofilament inclination increases; the peak transmitted force decreases as the knitted yarn inclination increases. This means that the knitted fabric with a higher knitted monofilament inclination and a longer knitted monofilament length more electively resists the impact due to a lower peak transmitted force [19, 20].

2.3 Bending Rigidity

Researchers have studied the bending properties of both warp- and weft- knitted fabrics. It appears that the bending rigidity of a knitted fabric is greatly related to the fabric type. Thus, a weft-knitted fabric has a higher bending rigidity in the weft-wise direction, while a warp-knitted fabric has a higher bending rigidity in the warp-wise direction. This behavior is mainly due to the directionality of the incorporated yarn [21]. When the samples are of the same fabric type (weft-knitted fabric, for example), we can further conclude that the bending rigidity is closely related to the fabric's density, knitted structure, and knitted type [22] They also found that weft-knitted fabrics using interlock structure, monofilament-knitted yarn, and a higher fabric density have a higher bending rigidity.

2.4 Stretch and Recovery

Researchers have found and suggested that the stretchability of the knitted fabrics is closely related to their fabric type [23]. The results from various reports reveal that the stretchability of a warp-knitted fabric has a high stretchability only in the weft-wise direction, while the stretchability in the warp-wise direction is very low (below 50%). On the other hand, weft-knitted fabrics have similar and high stretchability in both the weft-wise and warp-wise directions. As the knitted fabric is composed of two separate surface fabrics and linked together by a knitted yarn, it can be concluded that knitted fabrics carry the same fabric stretchability as their fabric types (i.e., warp-knitted or weft-knitted). When the results of the weft-knitted samples are compared, the stretchability in the weft-wise direction of samples is found to be higher than that in the warp-wise direction. This is due to using multifilament-knitted yarns, which have higher stretchability than those corresponding to samples using monofilament-knitted yarns [24].

2.5 Compressibility

Knitted fabrics are very resilient and resist and recover from pressure that may be applied on them; thus, deformation is not a problem in apparel made using knitted fabrics, and this may increase the life of the garment. Three distinct regions in this curve can be observed: modulus, collapse, and densification regions. The modulus of elasticity is defined as the initial slope in the linear elastic part of the stress–strain curve (modulus region). The initiation of collapse region is characterized by a relatively large deformation that occurs with a constant stress. During this stage, the monofilaments bend, so the thickness of the knitted fabric will decrease. This constant stress is referred to as a collapse stress or a collapse plateau. The most compressibility behavior and deformation of 3D fabrics occur in this region; thus, this region is the subject of many investigations in the cushion fabric mechanical behavior. In the densification region, monofilaments are engaged with each other, and the deflection change decreases; the slope of the stress–strain curve thus decreases [25–30].

2.6 Shear Properties

The shearing behavior of a fabric determines its performance properties when subjected to a wide variety of complex deformations during its use. The ability of a fabric to be deformed by shearing distinguishes it from other thin sheet materials such as paper or plastic films. This property enables the fabric to undergo complex deformations and to conform to the shape of the body [31]. Shear properties influence draping, flexibility, and handle of the fabric. The shear behavior of knitted fabrics was investigated by using a picture frame fixture. The image analysis procedure can provide much more information about the shear behavior of the fabric than stroke measurement. The displacement data and shear angles change during the loading process and can aid in the understanding of the shear behavior of the fabric. It is found that shear deformations depend very much on the type of knitted yarn and the fabric stitch density. The nonlinearity of shear deformation increases after limiting the locking angle that initiates the buckling of the sample [32].

2.7 Sound Absorption

The properties of knitted fabrics, such as 3D fiber disposition, possibility to use different materials, and single-step production system, enable them in different application areas. Researchers introduced fabric with knitted fabric structure to improve sound absorption performances. Their studies analyzed and reported that the acoustic performance of plain weft-knitted fabric is good in the middle- and high-frequency range [33]. Researchers analyzed and compared the effects of different fabric layers and arrangement sequences of both warp- and weft-knitted fabrics on the noise absorption coefficient [34]. They suggested that the sound absorption behavior of knitted fabrics is effective with multilayer arrangements backed up with air cavity. There are only few research studies conducted on the acoustic performance of knitted fabrics. Erhan Sancak determined that three factors have a major impact on the sound absorbance behavior: thickness of fabric, microporosity between fabric surfaces, and yarn linear density in the interconnection of the fabrics [35]. Researchers deeply discuss that the knitted fabrics have too much air in the pores, hence, sound energy dissipation may weaken when the porosity is higher than 0.9. The airflow resistivity is inversely proportional to the porosity of the fabrics; therefore, sound absorption increases with a decrease in porosity and increases with airflow resistivity. The knitted fabrics have a more tortuous path but a lower sound absorption because incident sound energy may get reflected away from the top layer and does not penetrate into the fabric. The thickness of the porous material layer has also a great influence on the position of the peak value in the frequency spectrum, but the effect of density is more predominant in terms of sound absorbency as compared to the effect of thickness [36].

2.8 Air Permeability and Moisture Management

Air permeability is another important factor that should be taken into account when choosing fabrics for certain applications. In this regard, weft-knitted fabrics have significantly better air permeability ratings and are thus more able to resist air penetration than warp-knitted fabrics [37]. However, it should be noted that the density of the fabric regardless of whether it is warp knit or weft knit will have a substantial impact on the air permeability and thermal regulation properties. A knitted fabric that is quite dense will have a higher thermal conductivity value but a low air permeability value; therefore, the end user must be taken into consideration to find an optimum density for the fabric [19, 38]. In regard to breathability, moisture wicking, and insulation of knitted fabrics, research at the Institute for Textile and Clothing Technology at the Hohenstein Institutes was conducted in regard to the insertion of a hydrophilic weft yarn on the face of the knitted fabric that is closest to the body, and its effect on the body's microclimate. A rib knitted construction can be made by using polyester (PES), monofilament for the pile and multifilament for the faces, and by various inserted weft yarns, which account for about 5% of the total fabric, in the face of the fabric closest to the body, which also had a ribbed construction. Researchers found that the inserted weft yarns increased the density of the fabric, and thus loaded the air permeability [39]. It was also found that viscose weft yarns increase heat transport by 14.2% when compared to knitted fabrics without inserted weft yarns. The inserted weft yarn, while increasing the thermal insulation properties of the fabric, can also increase the dimension stability of the fabric. Moisture management behavior is a vital factor in evaluating thermal and physiological comfort of functional textiles. Air and water vapor permeability, vertical wicking and moisture management of knitted fabrics with different fiber properties (fiber cross-section profile) have been studied quantitatively by various authors. They suggested that these knitted fabrics can be used for protective vest to absorb a user's sweat, to reduce the humidity, and improve user's thermal comfort [40]. For this reason, researchers have investigated the different 3D warp-knitted fabrics produced with functional fiber yarns in the back layer of the fabric (close to the body) and polyester in the front and middle layers (outer surface). Comfort properties such as air and water vapor permeability and wicking and other moisture management properties (MMP) of different fabric samples are measured. It is demonstrated that by using profiled fibers such as Coolmax fiber, the moisture management properties of knitted fabrics can be improved significantly. It is expected that these properties of 3D functional knitted fabrics, besides their good ventilation property, enable them to be used as a snug-fitting shirt worn under protective vests to feel lower humidity and to be dry and comfortable [41].

3 Quality Parameters Based on Applications of Knitted Fabrics

3.1 Cushioning Applications

Cushioning materials are used to dissipate the kinetic energy of the impacting mass while keeping the maximum load (or acceleration) below some limit [42]. They generally absorb kinetic mechanical energy under compression actions at relatively constant stress over a large range of displacement. The works done by compressing these kinds of materials are equivalent to the kinetic energies of a mass that might impact them. There are a number of materials and structures with the abovementioned feature for cushioning applications. Airbags, bubble films, rubberized bra cushioning, and polymer-based foams are just a few typical examples. However, despite their promising cushioning properties and low cost, the inferior comfort property makes these materials and structures unsuitable for human body protection. A combination of excellent transversal compressibility and high permeability makes knitted fabrics very suitable for multifunctional clothing and technical applications. Some efforts have already been made to investigate the compression properties of warp-knitted fabrics, and most of the studies have reported that the overall compression load–displacement relationship of these fabrics can be split into three main stages, i.e., linear elasticity, plastic plateau, and densification [42, 43]. This is the typical behavior required by a cushioning material in compression. In the published results, relatively constant loads at the plastic plateau stage are observed. These results have proved that warp-knitted fabrics are a new class of alternative candidate materials for cushioning applications. However, the plateau stage found in literature is not notable, and the zone of the plateau stage reported in the literature is also too short. In other words, the total energy absorbed in the plateau zone by these reported fabrics is not sufficient to identify them as good cushioning materials [44–47].

3.2 Knitted Fabrics for Composites

Knitted fabrics are three-dimensional textiles. Properties of knitted fabrics, such as 3D fiber location, possibility to use different materials, and production in one step, provide the knitted fabrics for use in different application areas. Due to inferior mechanical properties, such as elasticity and deformability under applied loads, conventional knitted fabrics are not suitable for high-performance composite applications. Moreover, the restricted distance between the plane layers contributes to the drawbacks of such knitted fabrics. One solution is to connect the planes by means of vertical fabric layers instead of pile yarns. This type of knitted fabric with multilayer reinforcements in the fabric structures is expected to show superior mechanical properties and be especially suitable as textile preforms for lightweight composite applications [48, 49]. Future applications of composites made from 3D multilayer knitted fabrics involve the replacement of conventional panel structures that are used for aircraft, transport vehicles, marine applications and infrastructures, lift cabins, and ballistic protection for buildings and combat vehicles. Researchers investigated knitted fabrics as reinforcement material in composite structures [50]. They also investigated the knitted fabric-reinforced composite materials produced with monofilament and multifilament knitted yarns. As a result of the study, it was concluded that multifilament knitted yarns provide better resin distribution although it is necessary to use monofilament knitted yarn to provide better compression resistance [48]. Researchers focused on the bending behavior of warp-knit knitted fabric–reinforced composite materials [51]. Others studied the application of knitted fabrics in composites [52]. Flat-knitted fabrics offer a strong potential for complex shape preforms, which could be used to manufacture composites with reduced waste and shorter production times. A reinforced knitted fabric made of individual surface layers and joined with connecting layers shows improved mechanical properties for lightweight applications such as textile-based sandwich preforms.

3.3 Protective Applications

Over the past few decades, a wide range of personnel protective equipment (PPE) has been developed to protect wearers from various types of risks or hazards to their health and safety [53]. Impact protectors, which are the most commonly used PPE, are normally manufactured to include energy-absorbing materials in the form of pads. They are integrated or inserted into protective clothing or equipment specially designed for protecting the human body from impact, blows, or falls. A number of different types of impact protectors are on the market for protecting different areas of the body in a variety of circumstances [54]. The use of warp-knitted fabrics in clothing and equipment providing protection against impact has attracted great attention in recent years due to their combination of protection and comfort in use. The static and dynamic compression behavior of a series of warp-knitted fabrics has been investigated in our previous studies, and the energy absorption performance and force attenuation capability of these fabrics under flatwise static and impact compression have been analyzed in detail [55]. These studies indicate that these fabrics have the key feature of behaving as cushioning materials, providing three distinct stages in static and dynamic compression, described as linear elasticity, plateau, and densification. However, in order to offer an adequate combination of protection and comfort, the protective material must conform to the shape and curvature of the body part being protected. There is no doubt that the impact properties of a protective material of a curved shape are different from those of a planar shape, due to the change in boundary conditions during loading. Most recently, Guo et al. have reported an experimental investigation into the impact behavior of warp-knitted fabrics of hemispherical shape [56]. The impact energy and weight of the striker are kept constant, and only the contact forces are measured. The tests are quoted as having been carried out according to the European Standard BS EN 1621–1:1998. However, this standard specifies that protectors should be impacted using a striker of 5 kg weight at a kinetic energy of 50 J, and the transmitted forces are then measured. Furthermore, their study did not pay attention to the relationship between energy absorption capacity and force attenuation properties, which is very important in designing fabrics to satisfy protective requirements.

3.4 *Knitted Fabrics for Thermo-physiological Clothing*

In the last few years, extensive research has been carried out on knitted fabrics for thermo-physiological comfort clothing [57, 58]. Knitted fabrics have the ability to trap and hold air and insulate the body because of the nature of knitted yarns between two surface layers. This, along with the ability to wick away moisture, maintains the body's microclimate, and thus keeps the person dry and comfortable. There are many outdoor/active apparel manufacturers who still employ the layering concept in order to achieve all the desirable properties in active apparel [59]. Warp-knitted fabrics tend to have a higher thermal insulation value than weft-knitted fabrics regardless of whether the fabric is wet or dry, an important feature for those who may utilize this fabric in the snow. Warp-knitted fabrics also have a higher thermal absorptivity value than weft-knitted fabrics, and thus the warp-knitted fabric will be warmer to touch than the weft-knitted fabric [60–63]. It was found that warp-knit fabrics have a lower thermal conductivity rating than weft-knit fabric, which means the excess heat from the body would not be as quickly transferred if a warp-knit knitted fabric is being utilized than if a weft-knit fabric is to be utilized with warp-knitted fabrics [64].

Knitted fabrics that have a rib construction, and are flocked during the finishing process, have been shown to provide better insulation and better heat and moisture transference, both of which will help maintain the body's microclimate [65]. The ribs of the face fabrics actually aid in the movement of heat and moisture; the channels created by ribbed fabric faces thus are better suited to control the microclimate and keep the wearer cool and dry. In regard to the water vapor permeability properties of knitted fabrics, weft-knitted fabrics have been found to have better evaporative heat loss properties and water vapor permeability properties than warp-knitted fabrics, thus making the weft knit more comfortable when worn close to the skin by a person who is exerting energy and perspiring. When choosing a fiber type to aid in microclimate regulation, viscose can absorb the perspiration of the wearer and delay the moment at which the air in the knitted layer becomes saturated with moisture [66]. This means that a person who is perspiring heavily, or has varying perspiration levels with high peaks, will remain drier for a longer period of time and have a more level microclimate, if wearing a garment utilizing knitted fabrics with a viscose-inserted weft yarn [67]. An inserted weft yarn, that is hydrophobic, will maintain the body's microclimate more effectively than an inserted weft yarn that is hydrophilic because the hydrophilic yarn holds the moisture and inhibits it from being transported away from the body [68]. Another, not commonly realized, important factor of knitted fabrics for certain types of activewear is compressibility. Many outdoor activities of athletes are seasonal (i.e., skiing), therefore, during the off-season, it is likely that the garments are stored in containers or they are under a heavy load. During the in-season, these garments are packed in suitcases for traveling. The athlete or outdoor enthusiast expects the garments to not be distorted in any way when they are removed from storage, as they should be ready to be utilized for the new season.

3.5 *Knitted Fabrics for Medical Applications*

As a 3D structure, the knitted fabric contains a considerable amount of space inside the fabric, and the knitted yarns oriented in the Z or thickness direction provide superior compression and recovery properties 24. In addition to the well-known advantages of knitted structures, such as high bursting strength, high elongation, low Young' s modulus, and high porosity, the knitted fabric stands out as a one-piece multilayered structure with high volume-to-weight ratio, softness, breathability, moisture conductivity, compression resistance, and excellent recovery properties [69–71]. This unique architecture with its impressive physical and mechanical performance has been discovered by medical textile researchers and applied to both internal and external end uses. In various external applications, the knitted layer allows consistent air circulation to reduce heat buildup and increase moisture transfer. Under applied pressure, it shows sustained graduated compression and uniform pressure distribution. So, it is ideally suited for use as a compression bandage, for comfort cushioning and shock absorbency [72–74]. When used as a compression bandage, for example, the knitted fabric provides lightweight, non-fraying, and breathable support with enhanced thermo-physiological properties and protective cushioning. In addition, knitted fabric shows excellent transference of pressure and the sub-bandage compression applied to the limb does not appear to be as severely influenced by the number of layers as it is with the traditional 2D bandages. For internal applications, knitted yarns can provide a layered surface area for cell attachment and guide cell migration through the thickness of the fabric. The numerous interconnecting pores allow fluids carrying nutrients and waste byproducts to flow through the entire structure, hence providing superior fluid transport performance. However, there is no published research data to support this claim at the present time. Another key advantage of using a knitted fabric as a tissue engineering scaffold is that it is a one-piece multilayer structure with different pore size distributions in the layers but without an interface. This means that its success fully avoids thermal bonding, toxic organic solvents, or chemical adhesives that would be used to assemble traditional foams and combine multilayer fabrics. The knitted fabric may also facilitate different cell lines so as to generate separate types of tissue in the different layers [75, 76]. The engineered pore size distribution and porosity gradient could also provide physical guidance for the differentiation of progenitor and stem cells. In summary, the future possibilities of applying knitted fabrics to a wide range of different tissue engineering end uses look promising, since various properties can be incorporated into the scaffold structure by changing the type of polymer, changing the type of yarns, modifying the manufacturing process or activating the fibers with a surface treatment. For instance, the total porosity and the pore size distribution can be controlled by a number of manufacturing parameters, such as the gauge of the needle bed, the number of yarn guides, the type or combination of yarns, the gap between the needle beds, and the orientation of the knitted yarns to parallel or crossed [77, 78].

4 Other Applications of Knitted Fabrics

The other technical applications of knitted fabrics and their products that are available in the market are shown below (Table 1) [79–82].

Since knitted fabrics have two outer surfaces connected to each other with knitted yarns, they provide lightweight and bulkier structure. So, the properties of knitted fabrics such as 3D fiber disposition, possibility to use different materials, and single-step production system enable them in different application areas. Components in knitted fabrics differ depending on the yarn type and production method. Excellent compression elasticity and breathability are the greatest advantages of knitted fabric. Admirable compressibility indicated that crush-resistant property and bending performance are excellent. Knitted fabric possesses excellent cushioning and shock-absorbing properties. This is because knitted fabric can absorb and dissipate kinetic mechanical energy when it is subjected to compression at regular stress over a large extent of displacement. The above discussion clearly

Table 1 Technical applications of knitted fabrics

Application fields	Knitted fabric products
Automotive	Car seat, door paneling, dashboard cover, car boot liners, car window shelf, car seat heating, etc.
Medical	Bandage, knee braces, thermal mats, wheelchair cushions, absorbent fleece, neck supports, etc.
Industrial	Textile antenna, sound absorption, solar thermal collectors, concrete reinforcement, etc.
Sports	Sports shoes, sports protectors, sportswear, mattress, pillows, etc.
Safety &protection	Cycle helmets, body armor, bulletproof jackets, hip protectors, etc.

stated the potential of knitted fabrics in a number of technical applications and holds promises for more applications. Also, knitted fabrics are still under active research for a number of advanced functional applications especially in mattresses, insole, automobile upholsteries, mats, etc.

5 Evaluation of Knitted Fabric Characteristics

5.1 Basic Constructional Parameters

Structural properties including the yarn linear density and fabric weights per unit area are determined according to the American Society for Testing and Materials (ASTM) D1059 standard using electronic weighing scales. The thickness of the fabrics was measured according to the ASTM D1777-96 standard with the SDL digital thickness gauge at a pressure of 200 Pa. The stitch density was calculated from wales per centimeter (WPC) and course per centimeter (CPC) with the help of an optical microscope. The density (D) of the fabric was calculated using the relationship in (Eq. 1)

$$D = \frac{W}{t} \mathrm{kg/m^3} \tag{1}$$

where W is the areal density (weight per unit area) and t is the thickness. Porosity, H, was calculated using Eq. 2:

$$H = 1 - \frac{\rho_a}{\rho_b} \tag{2}$$

where ρ_b is the bulk density of knitted fabrics and ρ_a is the weighted average absolute density of fibers in the knitted fabric, expressed in kg/m^3. All the experiments are carried out under standard ambient conditions and as per the standard testing method.

5.2 Compression Behavior

All the compression tests are carried out on a universal testing machine fitted with 5 kN load cell. The speed of compression was chosen at 12 mm/min in accordance with the ASTM d 575 standard (test methods for rubber properties). The compression test was performed on the machine equipped with two strictly parallel plates having a diameter of 150 mm and a smooth surface. The samples are cut with dimensions of 100 mm × 100 mm. All the knitted fabric specimens are compressed up to 80% of the initial thickness in an atmospheric condition of 20 °C and 65% relative humidity. Five tests are carried out for each sample under each testing condition, and the average compression stress–strain curves are presented.

5.3 Analysis of Compression Stress–Strain Curve of Knitted Fabrics

Normally, the compression behavior of knitted fabrics is classified into four stages with respect to changes in the slope. The four stages are (1) initial, (2) elastic, (3) plateau, and (4) densification (4). In the first stage, the surface layer of knitted fabric undergoes compression, a smaller slope is observed for loose/open structures, and slope increases with an increase in stitch density. The knitted yarns have a very low contribution in constraining the deformation during initial compression. Further compression (second stage) leads to a rapid increase in stress; it might be due to the jamming of surface yarns that allows monofilaments to buckle to a larger extent. In knitted fabrics, the third stage is quite complex because the compressive stress and strain are affected by buckling, shearing, and inter-contacting of knitted yarns. A faster increase in stress occurs in the fourth stage because the fabric achieves a very high density.

5.4 Energy Absorption During Compression of Knitted Fabrics

To understand the cushioning behavior of the knitted fabrics, it is necessary to evaluate and analyze the knitted fabric's energy-absorbing ability during compression. The compression curves reveal long deformation plateaus, suggesting that all knitted fabrics are potentially good energy-absorbing materials. The area under the load–displacement curve represents the total energy absorbed, and it can be calculated by multiplying the area under the stress–strain curve by the volume of the sample. The compression efficiency E is defined as the ratio of the energy absorbed by a real cushioning material compressed to a given strain and energy absorbed by an ideal cushioning material that transmits a constant stress of the same value at the same given strain [83, 84]. The compression stress, strain, energy absorption, and efficiency of both warp and weft-knitted fabrics are evaluated.

5.5 *In-plane Shear Behavior*

The in-plane shear behavior of both warp- and weft-knitted fabrics are carefully measured and analyzed using the picture frame test and image analysis methods.

5.5.1 Picture Frame Test

The picture frame shear test method has achieved some popularity in the early days of composite material development when few other shear test methods existed. However, as the two- and three-rail shear test methods and later, the Iosipescu shear test were introduced for characterizing basic shear properties, picture frame shear test became less popular for three reasons: It used a relatively large specimen; test preparation required that a number of holes be drilled in the specimen; and the method required a complex fixture. Despite these disadvantages, the picture frame shear test continued to be an attractive option for composite laminate panel testing because the method accommodates large specimens [85].

As shown in Fig. 1, the picture frame is an effective way for characterizing the intra-ply shear property of fabrics. The picture frame test is preferred by many researchers for shear testing since it has a pure state of strain that can be imposed on the test specimen. Shearing is induced by restraining the textile reinforcement in a rhomboid deformation frame with fibers constrained to move parallel to the frame edges. The frame is extended at diagonally opposing corners using simple tensile testing equipment. The description of this test method and the modifications done in the frame for this work have been discussed in detail in the following section.

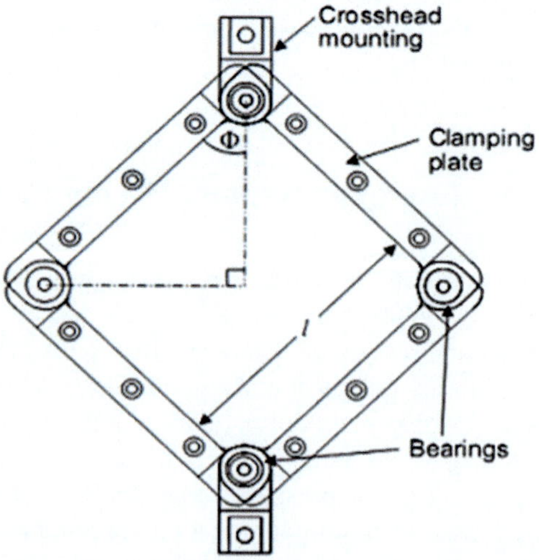

Fig. 1 Picture frame fixture design

5.5.2 Description of the Test Method

The apparatus has four legs that are hinged to form a picture frame. Further, there is a rod that runs across one of the diagonals of the frame. It should be noted that this rod is not in the plane of the frame but runs behind the frame. The lower end of the rod is hinged with the common hinge of the two legs meeting in that corner. The other opposite corner hinge rests in the slot provided in the rod. The slot is about 6 cm long for maximum deformation of the frame. Further, at the lower end, the rod is again hinged to individual legs of the frame. The lower end of the rod is fixed in the crossheads of the loading machine. By adjusting the distance between the upper and lower crossheads, the angles between the arms of the fixture reach 90°.

The distance can be set as the original reference value, i.e., the zero displacement position, in the computer, so that later on, all the experiments can automatically begin from this zero point. Also, the force can be set to zero at this position. Thus, when the load is applied through the upper end of the rod, it pushes these two legs apart and deforms the frame. These two legs in turn push their adjacent legs making their common hinge slide in the slot of the rod. The plate deforms into a diamond shape. The knitted fabric is clamped to the frame with the help of clamping plates. These clamping plates are 3 cm in width and have a diamond knurled surface for better gripping. The empty frame is tested under the same condition to find the frictional effects between bearings and slots, after several trials of this test; the average value of load at each displacement point is calculated. This is to record the load–displacement behavior of the empty fixture under the same condition as in the real shear experiment. So, it is mainly considered and optimized during sample testing. To eliminate the error caused by the weight and inertia of the fixture, the net load obtained was subtracted from the machine-recorded load when the fabric was deformed in the picture frame. This resultant load is considered and accounted for as an actual load. In order to prevent the pressures from being imposed on test samples by the fixture during the large deformation, the central area of shear deformation was 100 mm × 100 mm, and the four corner parts are cut off in order to avoid edge buckling. Shear tests are conducted on a universal tensile testing machine with a crosshead speed of 10 mm/min. The test was repeated for five samples of each type under the same conditions [86].

5.6 *Thermo-physiological Properties*

The air, heat, and water transmission behavior of both warp- and weft-knitted fabrics are measured and analyzed for advanced applications.

5.6.1 Air Permeability

Air permeability is described as the rate of airflow passing perpendicularly through a known area, under a prescribed air pressure differential between the two surfaces of a material. Tests are performed according to standard ISO 9237 using a Textest FX-3300 air permeability tester. The air pressure differential between the two surfaces of the material was 200 Pa.

5.6.2 Thermal Properties

The measurements of thermal insulation parameters are performed on knitted fabrics with the use of the ALAMBETA device constructed in the Czech Republic. Alambeta measuring device was used for fast evaluation of transient and steady-state thermo-physiological properties (thermal insulation and thermal contact properties). The instrument also measures the sample thickness. The instrument consists of two measuring heads between which the test specimen was placed. Both measuring heads are equipped with thermocouples and heat flow sensors. The lower measuring head was adjusted to the ambient temperature by suitable cooling means; the upper, heated measuring head was adjusted to a controlled constant differential temperature. The heat flow sensors act at the contact faces of both measuring heads. When the upper measuring head was loaded on the measuring specimen, the heat flow at the upper surface and the underside of the test specimen could be measured. The fundamental measuring principle implies the measuring and processing of the heat flow in dependence of time. Six parameters are determined: thermal conductivity λ, thermal diffusion a, thermal absorption b, thermal resistance r, the ratio of maximal to stationary heat flow density υ, and stationary heat flow density q_s at the contact point.

5.6.3 Water Vapor Permeability

The water vapor permeability of the samples is measured using the PERMETEST. The instrument works on the principle of heat flux sensing. The fabric sample is placed on a measuring head over a semipermeable foil and exposed to parallel airflow at a velocity of 1 m/s. The temperature of the measuring head is maintained at room temperature for isothermal conditions. When water flows into the measuring head, some amount of heat is lost. This instrument measures the heat loss from the measuring head due to the evaporation of water in bare condition and while being covered by the fabric. The relative water vapor permeability (*RWVP*) of the fabric sample is calculated by the ratio of heat loss from the measuring head with fabric (q_s) to that without fabric (q_o) as given in Eq. (3).

$$RWVP = \frac{q_s}{q_o} \times 100\% \tag{3}$$

PERMETEST characterizes the capability of the fabric to transfer water vapor, by measuring two parameters, the relative water vapor permeability and the absolute evaporative resistance (*Ret*).

6 Acoustic Properties

6.1 *Determination of Tortuosity*

Tortuosity is a fundamental parameter that describes the complexity of the path of a sound wave propagating within a porous material. There are a number of methods that have been developed to measure tortuosity. It is mainly based on high-frequency acoustic transmission, ultrasonic pulse transmission, reflection, and electric resistivity measurements. In this work, the tortuosity of knitted fabrics was determined by both experimental method using ultrasonic waves and analytical method.

6.2 *Airflow Resistance*

As stated earlier, the viscous resistance of air in the porous material has an important influence on the sound absorption mechanism. Flow resistance has been used as an important parameter in theoretical equations by many researchers. It is therefore important to measure the flow resistance of an acoustic sample.

Airflow resistance of knitted fabric was calculated from the air permeability value obtained from Textest FX-3300 air permeability tester. The air permeability is described as the rate of airflow passing perpendicularly through a known area, under a prescribed air pressure differential between the two surfaces of a material (200 Pa). Tests are performed according to standard ISO 9237 for five specimens of each sample and expressed as linear airflow velocity (v) in m/s. This specific flow resistance has been converted to airflow resistivity, *R* using Eq. (4) shown below.

$$\text{Airflow Resistivity}, R = \frac{\nabla P}{v^* d} \text{Pa.m}^{-2}\text{.s} \tag{4}$$

where R is the airflow resistivity,
∇P is the pressure difference in Pascal,
v is the linear airflow velocity in m/s, and,
d is the thickness in meter.

6.3 Sound Absorption Properties

The frequency of resonance depends on the ratio of elastic modulus to density as both these parameters can be varied independently. Knitted fabrics can be used to eliminate certain frequencies. The damping capacity of knitted fabrics has been shown to be an order of magnitude higher than that of other nonwoven fabrics and foams. Knitted fabrics have also been used as soundproofing materials. Within porous and three-layer structures, sound is attenuated by vibration and friction losses as gas flows between two layers during propagation. Repeated reflections within the knitted structure give rise to long paths where full absorption is possible. The acoustic properties of the knitted fabrics can be used where sound absorption is vital, i.e., auditoriums, automotive vehicles, etc. There are several techniques to measure sound absorption: reverberant field method, impedance tube method, and steady-state method [86, 87].

6.4 Measurement of Sound Absorption Coefficient (Impedance Tube Method)

In this research, the impedance tube method was used to determine the normal incident sound absorption coefficient, SAC (α). A minimum of three specimens for each sample are tested according to ASTM E 1050–07. A standard test method for impedance and absorption of acoustic materials using a tube with two microphones and a digital frequency analysis system were used (Fig. 2). It uses plane sound waves that strike the material straight, and so the sound absorption coefficient is called normal incidence sound absorption coefficient, SAC. In this study, the impedance tube method was used, which is faster and generally reproducible and, in particular, it

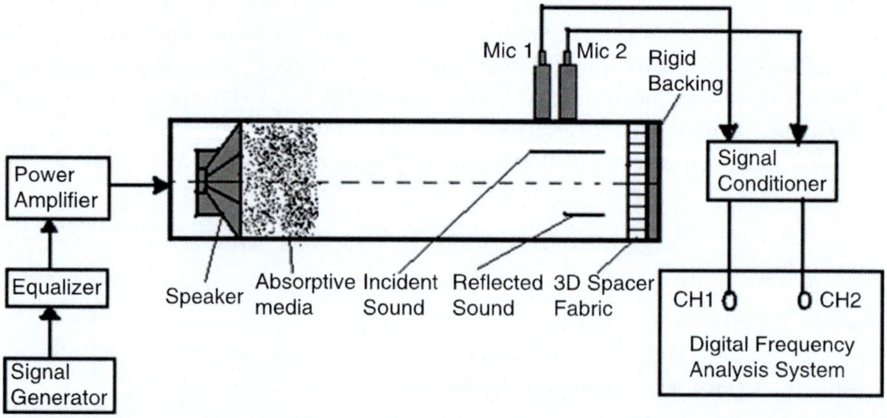

Fig. 2 Impedance tube method

requires relatively small circular samples, both 29 and 100 mm in diameter according to the frequency range (former measures 500 Hz–6.4 kHz and later 50–500 Hz). Thus, the method avoids the need to fabricate large test samples with lateral dimensions several times the acoustic wavelength.

6.5 *Calculation of Noise Reduction Coefficient*

The noise reduction coefficient (NRC) is a measure of how much sound is absorbed by a particular material and is derived from the measured sound absorption coefficients at different frequencies (Hertz). The NRC was determined using the following formula (Eq. 5).

$$\mathrm{NRC} = \frac{\alpha_{250} + \alpha_{500} + \alpha_{1000} + \alpha_{1500} + \alpha_{2000} + \alpha_{3000} + \alpha_{4000} + \alpha_{5000} + \alpha_{6000} + \alpha_{6400}}{10} \quad (5)$$

7 Statistical Analysis

Statistical analysis software are extensively used to conduct all the statistical tests. Advance statistical evaluation and two-way analysis of variance are used to analyze the significance of various factors on the required properties of warp- as well as weft-knitted fabrics. Also, differences in means between various groups of fabrics can be examined for statistical significance using one-way ANOVA followed by pair comparison using theoretical methods. For all the statistical tests, differences are considered significant at $P < 0.05$. Data are reported as mean ± standard error of mean, unless otherwise stated.

References

1. W. Fung, M. Hardcastle, Product engineering – Interior trim, textiles in automotive engineering, in *Textiles in Automotive Engineering*. (The Textile Institute/Woodhead Publishing, Cambridge, 2001) ISBN 1 85573 493 1, pp. 194–211
2. R. Deng, P. Davies, K. Bajaj, Flexible polyurethane foam modelling and identification of viscoelastic parameters for automotive seating applications. J. Sound Vib. **262**(3), 391–417 (2003)
3. L.J. Gibson, M.F. Ashby, *Cellular Solid Structure and Properties* (Pergamon Press, New York, 1988)
4. M. Pajon, M. Bakacha D. Pignede, P. Van Effenterre, in *Modeling of P.U. Foam Behavior – Applications in the Field of Automotive Seats*, Bertrand Faure Equipments S.A., SAE 960513, International Congress and Exposition Detroit, Michigan, Feb 1996
5. G.R. Blair, A. Milivojevich, J.D. Van Heumen, *Automotive Seating Comfort: Investigating the Polyurethane Foam Contribution – Phase I*, SAE Paper No. 980656, Detroit (1997)

6. K. Ebe, *Effect of Thickness on Static and Dynamic Characteristics of Polyurethane Foam* (International Congress of Noise Control Engineering, Yokohama, 1994)
7. M. Kolich, S.D. Essenmacher, J.T. McEvoy, Automotive seating. The effect o f foam physical properties on occupied vertical vibration transmissibility. J. Sound Vib. **281**(1–2), 409–416 (2005)
8. J. Hopkins, A comparative analysis of laminating automotive textiles to foam. J. Coat. Fabr., 250–267 (1995)
9. R.A. Pfeifer, Evaluation of textile fabric properties utilized in foam-in-place head restraints. Master Thesis, Eastern Michigan University, 12 Oct 2005
10. N. Njeugna, L. Schacher, D.C. Adolphe, J.B. Schaffhauser, P. Strehle, Development of a new 3D nonwoven for automotive trim applications, in *New Trends and Developments in Automotive Industry* (InTech, 2011, ISBN: 978-953-307-999-8), pp. 323–346
11. X. Wu, S. Rakheja, P. Boileau, Distribution of human-seat interface pressure on a soft automotive seat under vertical vibration. Int. J. Ind. Ergon. **24**, 545–557 (1999)
12. I. Hostens, G. Papajoannou, A. Spaepen, H. Ramon, Buttock and back pressure distribution tests on seats of mobile agricultural machinery. Appl. Ergon. **32**(4), 347–355 (2001)
13. T.G. Cengiz, F.C. Bablik, In on -the-road experiment into the thermal comfort of car seats. Appl. Ergon. **38**(3), 337–347 (2007)
14. N.C. Hilyard, P. Collier, Effect of vehicle seat cushion material on ride comfort. Paper presented at plastic on the road, the plastic and rubber institute international conference, London, 1984
15. Y. Liu, H. Hu, Compression property and air permeability of weft-knitted spacer fabrics. J. Text. Inst. **102**(44), 366–372 (2011)
16. Y. Liu, L. Lv, B. Sun, H. Hu, B. Gu, Dynamic response of 3d biaxial spacer weft-knitted composite under transverse impact. J. Reinf. Plast. Compos. **25**(15), 1629–1641 (2006)
17. Y. Liu, H. Hu, L. Zhao, H. Long, Compression behavior of warp-knitted spacer fabrics for cushioning applications. Text. Res. J. **82**(1), 11–20 (2012)
18. M. Li, S. Wang, Z. Zhang, B. Wu, Effect of structure on the mechanical behaviors of three-dimensional spacer fabric composites. Appl. Compos. Mater. **16**, 1–14 (2008)
19. J. Yip, S. Ng, Study of three-dimensional spacer fabrics: Physical and mechanical properties. J. Mater. Process. Technol. **206**(1–3), 359–364 (2008)
20. D. Gross, 3D spacer knit fabrics for medical applications. J. Text. Appar. Technol. Manage. **4**, 26–28 (2003)
21. D. Ionesi, R. Ciobanu, A. Vircan, M. Blaga, C. Budulan, Three – Dimensional knitted fabric with technical destination, Universitatea Tehnică "Gheorghe Asachi" din Iaşi Tomul LVI (LX), Fasc. 3, 2010, pp. 29–37
22. L. Ciobanu, Development of 3D knitted fabrics for advanced composite materials, in *Advances in Composite Materials – Ecodesign and Analysis*, ed.by B. Attaf (InTech, 2011). ISBN: 978-953-307-150-3
23. S.H. Bruer, N. Powel, G. Smith, Three dimensionally knit spacer fabrics: A review of production techniques and applications. J. Text. Appar. Technol. Manage. **4**, 1–31 (2005)
24. K.D. Darlington, The productions of Raschel crochet fabric, Knit Outwear Times, 1968, vol. 22, pp. 42–45
25. B. Wheatley, Development of tricot and Raschel machinery over the past 50 years. Knit Outwear. Times Y'r Bk, 1968, pp. 242–257
26. http://www.textileworld.com/textile-world/knitting-apparel/2015/03/the-wide-world-of-knits/
27. T. Palanirajan, G. Ramakrishnan, S. Sundaresan, P. Kandhavadivu, The influence of fabric parameter on low-stress mechanical properties of polyester warp-knitted spacer fabric. Int. J. Fash. Des. Technol. Educ. (2016). https://doi.org/10.1080/17543266.2016.1177738
28. Y. Liu, H. Hu, Three-dimensional knitted textiles, in *Advances in 3D Textiles*, ed. by X. Chen, (The Textile Institute/Wood Head Publishing, Cambridge, 2015), pp. 125–152
29. W. Fung, M. Hardcastle, *Textile and Automotive Engineering* (The Textile Institute/Woodhead Publishing, 2001)
30. W. Fung, *Coated and Laminated Textile* (The Textile Institute/Woodhead Publishing, 2002)

31. Y. Liu, H. Hu, in *Vibration Isolation Performance of Warp-knitted Spacer Fabrics*. Fiber Society Spring Conference, Hong kong, 23–25 May 2011, pp. 63–64
32. G. Smith, Industrial fabric products review. Buyers Guide **60**(10), 42–44 (2004)
33. C. Iyer, B. Mammel, W. Schach, *Circular Knitting* (Meisenbach, 1992)
34. G. Ertekin, A. Marmarali, Heat, air and water vapor transfer properties of circular knitted spacer fabric. Textil ve Konfeksiyon **4**, 369–373 (2011)
35. M. Abounaim, Process development for the manufacturing of flat knitted innovative 3D spacer fabrics for high performance composite applications. PhD. Thesis, Technical University of Dresden, 2011
36. M. De Araujo, R. Fangueiro, A. Cataring, H. Hong, Recent developments in weft – Knitting science and technology: The way a head in the new millennium. Revista Romana de Textile – Pielarie **1**, 61–66 (2001) ISSN1453-5424
37. Y.P. Liu, H. Hu, H. Long, L. Zhao, Compression behavior of warp-knitted spacer fabrics for cushioning application. Text. Res. J. **82**, 21–26 (2012)
38. Y.P. Liu, H. Hu, H. Long, L. Zhao, Impact compressive behavior of warp-knitted spacer fabrics for protective applications. Text. Res. J. **82**, 773–288 (2012)
39. Y. Liu, H. Hu, W.M. Au, Effect of structural parameters and lamination protective properties of warp-knitted spacer fabrics under impact in hemispherical form. Part II. Text. Res. J. (2013). https://doi.org/10.1177/0040517513495942
40. K. Machova, P. Klug, M. Waldmann, G. Hoftmann, C. Cherif, Determining of the bending strength of knitted spacer fabric. Melliand Textilberichte **87**(6), E93 (2006)
41. V. Arumugam, R. Mishra, J. Militky, M. Tunak, In-plane shear behavior of 3d knitted spacer fabrics. J. Ind. Text. **46**, 868–886 (2016)
42. M. Xu-hong, G. Ming-qiao, The compression behaviour of the warp knitted spacer fabric. Fibers Text. East. Eur. **16**(1), 56–61 (2008)
43. V. Arumugam, R. Mishra, J. Militky, J. Salcova, Investigation on thermo-physiological and compression characteristics of weft knitted spacer fabrics. *J. Text. Inst.*, Epub ahead of print, 12 Aug 2016. https://doi.org/10.1080/00405000.2016.1220035
44. T. Dias, R. Monaragala, P. Needham, E. Lay, Analysis of sound absorption of tuck spacer fabrics to reduce auto-motive noise. Meas. Sci. Technol. **18**(8), 2657–2666 (2007)
45. Y. Liu, H. Hu, Sound absorption behavior of knitted spacer fabrics. Text. Res. J. **80**(18), 1949–1947 (2010)
46. E. Sancak, An investigations of sound absorbance properties of weft knitted spacer fabrics. Int. J. Acoust. Vib. **20**(1), 36–40 (2015)
47. V. Arumugam, R. Mishra, J. Militky, J. Novak, Thermo – Acoustic behavior of 3D knitted spacer fabrics. Fibers Polym. **16**(11), 2467–2476 (2015)
48. X. Ye, H. Hu, X. Feng, Development of the warp knitted spacer fabrics for cushion applications. J. Ind. Text. **37**(3), 213–223 (2008)
49. K. Machova, G. Hoffmann, C. Cherif, Micro-climate regulation of clothing systems. Melliand Textilberichte Internat Textile Reports German Incl Yellow Engl Tra **87**, 11–12 (2006)
50. M. Bagherzadeh, M. Gorji, P. Latifi, Payvandy, L.X. Kong, Evaluation of moisture management behavior of high-wicking 3D warp knitted spacer fabrics. Fibers Polym. **13**(4), 529–534 (2012)
51. S.C. Anand, in *Recent advances in knitting technology and knitted structures for technical textiles applications*. Proceedings from ISTEK, 2003
52. X. Ye, R. Fangueiro, H. Hu, M. Araujo, Application of warp-knitted spacer fabrics in car seats. J. Text. Inst. **98**, 337–344 (2007)
53. D.M. Armakan, A. Roye, A study on the compression behavior of spacer fabrics designed for concrete applications. Fibers Polym. **10**, 116–123 (2009)
54. D. Mecit, A. Roye, Investigation of a testing method for compression behavior of spacer fabrics designed for concrete applications. Text. Res. J. **79**, 867–875 (2009)
55. S. Savci, J.I. Curiskis, M. Pailthorpe, Knittability of glass fiber weft-knitted preforms for composites. Text. Res. J. **71**, 15–21 (2001)

56. M. Abounaim, G. Hoffmann, O. Diestel, C. Cherif, in *3D Spacer Fabric as Sandwich Structure by Flat Knitting for Composites Using Hybrid Yarn*. Proceedings of the Autex Conference, Izmir, 2009, pp. 675–681
57. I. Verpoest, *Characterisation and Development of 3D Knitted Composites* (Katholieke 78 Universitaet Leuven, Department of Material and Metallurgy Engineering Composite Materials Group, 1995)
58. D. Phillips, I. Verpoest, J. van Raemdonck, in *Optimising the Mechanical Properties of 3D-Knitted Sandwich Structures*. Proceedings of the 11th International Conference on Composite Materials, 14–18 July 1997, pp. 211–218
59. D. Philips, I. Verpoest, in *3D-Knitted Sandwich Structures: Production, Properties and Prospects*. Proceedings ECCM-8, 8th European Conference on Composite Materials – Science, Technologies and Applications, vol. 1, Italy, 3–6 June 1998, pp. 543–550
60. D. Mecit, A. Marmarali, Application possibilities of 3d weft knitted spacer fabrics in composite structures, AUTEX 2011, 8–10 June 2011, Mulhouse
61. A.S. Richard, *Textiles for Protection* (The Textile Institute/Woodhead Publishing, Cambridge, 2005) ISBN 1 85573 493 1
62. D. Rothe, Warp knitted s pacer fabric – Design and application fields. Knitting Technol. **4**, 14–16 (2001)
63. Y. Liu, W.M. Au, H. Hu, Protective properties of warp-knitted spacer fabrics under impact in hemispherical form. Part I: Impact behavior analysis of a typical spacer fabric. Text. Res. J. **84**(4), 422–434 (2014)
64. X.F. Guo, H.R. Long, L. Zhao, Investigation on the impact and compression-after-impact properties of warp-knitted spacer fabrics. Text. Res. J. **83**, 904–916 (2013)
65. N. Oglakcioglu, A. Marmarali, Thermal comfort properties of some knitted structures. Fibers Text. East. Eur. **15**(5–6), 64–65 (2007)
66. N. Ucar, T. Yilmaz, Thermal properties of 1x1, 2x2, 3x3 rib knit fabrics. Fibers Text. East. Eur. **3**(47), 34–38 (2004)
67. B. Crina, M. Blaga, V. Luminita, R. Mishra, Comfort properties of functional weft knitted spacer fabrics. TekstiL Ve Konfeksiyon **23**(2), 220–227 (2013)
68. B. Pause, Thermo-physiological comfort provided by knitted spacer fabrics. Melliand Textileberichte **83**(3), 134–136 (2002)
69. V. Krel, G. Hoffmann, P. Offermann, K. Machova, L. Hes, Spacer fabrics for sports clothing with improved comfort. Melliand Textilberichte **86**(5), E73 (2005)
70. S. Anand, C. Rebenciuc, in *Elaboration of a Prediction Method of the Values for Some Characteristics of the Weft Knitted Fabrics*. 5th International Conference TEXSCI 2003 Proceedings, Liberec, 2003
71. A. Reisfeld, *Warp Knit Engineering* (National Knitted Outerwear Association, New York, 1990)
72. S. Rajendran, S. Anand, Design and development of novel bandages for compression therapy. Br. J. Nurs. **12**(6), S20–S29 (2003)
73. A. Davies, J. Williams, The use of spacer fabrics for absorbent medical applications. J. Fiber Bioeng. Inform. **1**(4), 321–329 (2009)
74. X.H. Miao, M.Q. Ge, The compression behavior of warp knitted spacer fabric. Fibers Text. East. Eur. **16**, 90–92 (2008)
75. T. He, A study of three dimensional warp knits for novel applications as tissue engineering scaffolds. A Master Thesis, NC State University, 2011
76. G. Lee, S. Rajendran, S. Anand, New single-layer compression bandage system for chronic venous leg ulcers. Br. J. Nurs. **18**(15), S4–S18 (2009)
77. F.T. Moutos, B.T. Estes, F. Guilak, Multifunctional hybrid three-dimensionally woven scaffolds for cartilage tissue engineering. Macromol. Biosci. **10**(11), 1355–1364 (2010)
78. F.T. Moutos, L.E. Freed, F. Guilak, A biomimetic three-dimensional woven composite scaffold for functional tissue engineering of cartilage. Nat. Materials **6**(2), 162–167 (2007)
79. M. Rock, K. Lohmueller, Three dimensional knit spacer fabric for bed pads. US patent No. 5817391, 1998

80. S. Bawadi, Development of the weaving machine and 3D woven spacer fabric structures for lightweight composites materials. PhD Dissertation, Technische Universität Dresden, 2007
81. A. Unal, G. Hoffmann, C. Cherif, Development of weft knitted spacer fabrics for composite materials. Melliand Textileberichte **4**, 49–50 (2006)
82. M. Abounaim, G. Hoffmann, O. Diestel, C. Cherif, Thermoplastic composite from innovative flat knitted 3D multi-layer spacer fabric using hybrid yarn and the study of 2D mechanical properties. Compos. Sci. Technol. **70**, 363–370 (2010)
83. L. Gilbert, N.B. Martin, D. Johanne, Evaluation of bias-extension and picture-frame test methods for the measurement of intra ply shear properties of PP/glass commingled fabrics. Compos. Struct. **61**, 341–352 (2003)
84. A. Willems, S. Lomov, I. Verpoest, D. Vandepitte, in *Picture Frame Shear Tests on Woven, Textile Composite Reinforcements with Controlled Pretension*. 10th Esa form conference on material forming, Parts A and B, 2007, pp. 999–1004
85. M. Avalle, G. Belingardi, R. Montanini, Characterization of polymeric structural foams under compressive impact loading by means of energy-absorption diagram. Int. J. Impact Eng. **25**, 455–472 (2001)
86. P.H. Mareze, R.P. Becker, A. Lenzi, C. Pellegrini, in *Rigid-Frame Porous Material Acoustic Attenuation on Compressor Discharge*. International Compressor Engineering Conference at Purdue July 16–19, 2012
87. F. Fohr, D. Parmentier, B. Castagnede, M. Henry, An alternative and industrial method using low frequency ultrasound enabling to measure quickly tortuosity and viscous characteristic length. Proc. Acoust. (2008)

Technical Applications of Knitted Fabrics

Rajesh Mishsra and Hafsa Jamshaid

1 Medical Applications

Medical textiles are made with woven, nonwoven, braided and knitted fabrics. The knitted medical textiles are rather technical textiles including artificial blood vessels, hernia patches, compression bandages, vascular prostheses, surgical hosiery, blankets, wound dressings, stockings, elasticated net structures, expandable metallic stents, etc. The knitted medical textiles have replaced in large numbers the synthetic medical products made of rubber, foam, latex and other laminated structures. The knitted medical textiles are made with both warp and weft knitting. The reason behind the preference for knitting is high flexibility, air permeability, adaptivity and free motion capabilities of knitted structures [1]. The medical textiles are divided into different categories including non-implantable, implantable, extra-corporeal, healthcare and hygiene products. The weft-knitted structures which are mostly used for medical textile applications are plain and rib structures as shown in Fig. 1.

The warp-knitted structures which are used for medical applications are tricot, atlas, locknit and pillar stitches as shown in Fig. 2. These structures can be modified or combined with other structures as per requirement.

There is an increasing trend to use 3D spacer fabric (Fig. 3) for medical applications these days due to its unique structure and properties. The spacer fabrics can be produced both on warp and weft knitting machines, but the 3D spacer fabrics

R. Mishsra
Faculty of Engineering, Czech University of Life Sciences Prague, Prague, Czech Republic

H. Jamshaid (✉)
School of Engineering and Technology, National Textile University, Faisalabad, Pakistan
e-mail: hafsa@ntu.edu.pk

H. Jamshaid, R. Mishra (eds.), *Knitting Science, Technology, Process and Materials*, Textile Science and Clothing Technology,
https://doi.org/10.1007/978-3-031-44927-7_8

Fig. 1 (**a**) Plain single-knit structure (**b**) rib double-knit structure

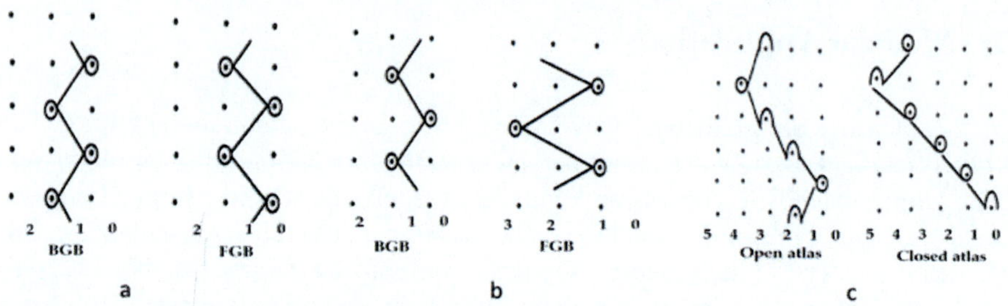

Fig. 2 Lapping movements of (**a**) tricot structures (**b**) locknit structure (**c**) open and close atlas [2]

produced on the weft knitting (double cylinder) machines are more cost-effective as there is no need to prepare the beams like warp knitting. The spun and coarser yarn count can also be used for weft-knitted spacer fabrics.

1.1 Non-implantable Medical Textiles

The non-implantable medical textiles mostly include products which are used to cover wounds and help in healing, e.g., medical dressing, gauze, bandages, etc.

The purpose of medical dressing is to cover the wound, insulate, transfer the drugs and absorb liquids. When it comes to knitted wound dressings, they are permeable, elastic and fit to the body, hence can be used with greater ease. The knitted dressings also hinder the muscles less during body movement. Both warp- and weft-knitted fabrics are used in wound dressings. Mainly weft-knitted fabrics are used for wound dressings with plain and rib structures (Fig. 1). Cotton, viscose filament, alginate fiber, jute cell, bamboo and chitosan are used to make medical dressings [3].

Bandages are used to provide support to medical devices or patient body. Warp-knitted medical bandages are used, which are made with pillar and tricot stitches. The knitted bandages include high-support bandages, compression bandages, orthopedic bandages and elastic bandages. Medical gauzes are also made with knitted structures [3].

Fig. 3 Weft-knitted 3D spacer fabric illustration

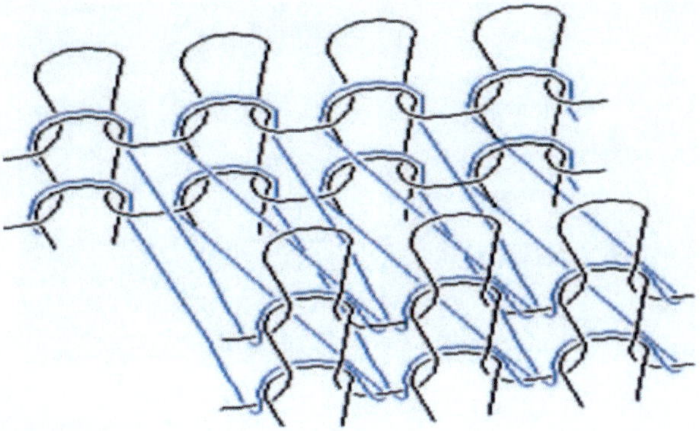

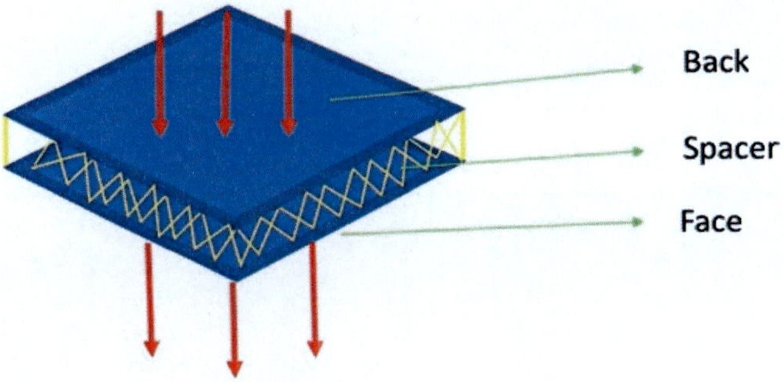

Fig. 4 Moisture and heat transfer phenomenon in spacer fabrics

3D knitted bandages are also used due to their good moisture and heat management properties. In the 3D spacer structures, different types of materials and structure combination can be used in any layer. The 3D spacer fabric bandages for venous leg ulcer offer better elasticity and improved healing [1]. Figure 4 shows the mechanism of moisture and heat transfer in 3D spacer fabrics.

Cotton, viscose, lyocell, polyamide, elastomeric-fiber yarn, glass and water-soluble Polyvinyl alcohol (PVA) are used in gauzes and medical bandages [3, 4].

1.2 Implantable Medical Textiles

Implantable medical textiles are implanted in the human body, so they must have biocompatibility and flexibility to be adjusted with the body. These include artificial blood vessels, artificial ligaments, hernia patches, etc.

Warp-knit atlas and tricot stitches are used to make artificial blood vessels (Fig. 5). Polyester, polypropylene, polyethylene, polyurethane, polytetrafluoroethylene and some other synthetic fibers are used to make artificial blood vessels.

Artificial ligament is used to replace damaged knee joints. The artificial ligaments are made in warp knitting machines using polyester fiber. Hernia mesh (Fig. 5b, c) is used to prepare the hernia and made with warp-knit structures using pillar stitch, tricot stitch and atlas stitch (Fig. 2). Polyester, polypropylene, expanded

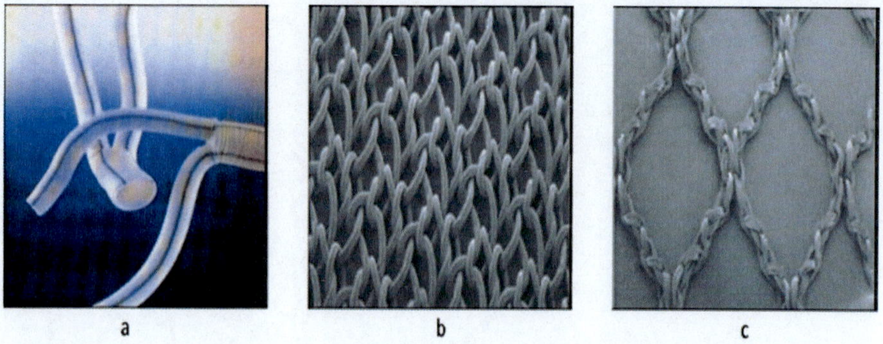

Fig. 5 Warp-knitted (**a**) artificial blood vessel (**b**) Marlex® patch (Hernia patch) (**c**) Vypro® patch (Hernia patch) [3]

Fig. 6 Implantable cardio support devices [5]

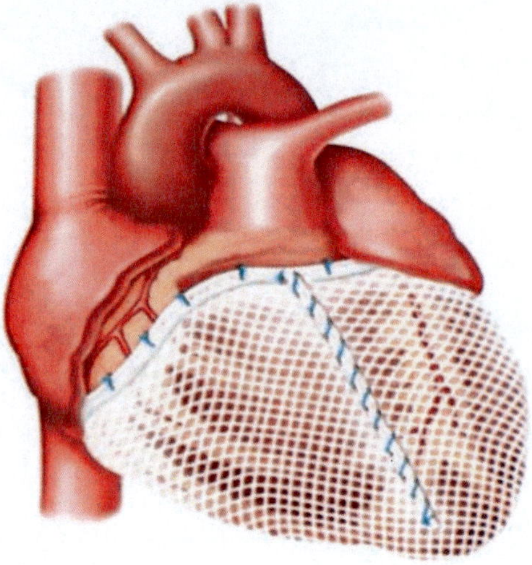

polytetrafluoroethylene, polyvinylidene fluoride and absorbable polymers such as polyglycolic acid are used to make hernia mesh.

Cardiac support devices (Fig. 6) made with warp-knitted atlas stitches are used to support human heart for remodeling of heart muscles. Acorn Cardiovascular Inc. has used polyester multifilament to make cardiac support devices [5].

Knitted expandable metallic stent can also be used depending upon the application. The knitted expandable metallic stent (Fig. 7) was made with nitinol (55% nickel and 45% titanium) by Boston Scientific. The expandable metallic is transferred through airways by flexible or rigid bronchoscopy, so it can pass by tumor easily [6].

Both warp and weft knitting techniques are used to make tendon scaffold, but plain weft-knit structures are preferred over warp-knit structures. The knitted

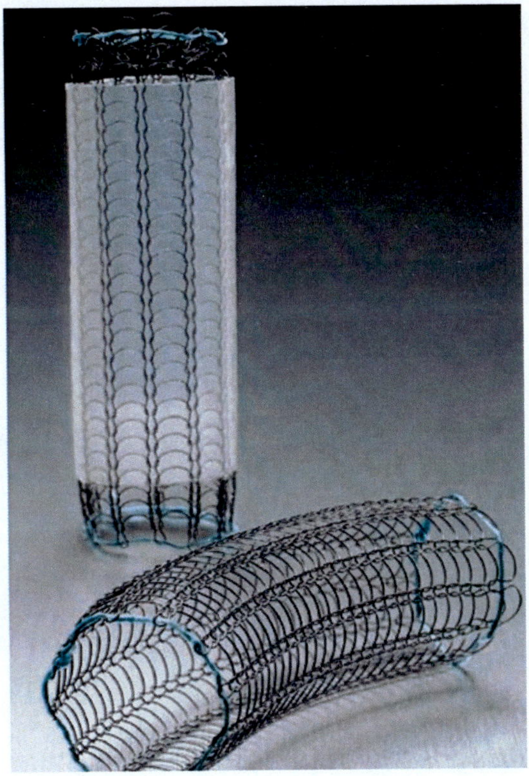

Fig. 7 Ultraflex™ knitted metallic stent by Boston Scientific [6]

scaffold offers low strength than conventional scaffold due to its intermeshed structure but offers a greater possibility of tissue regeneration. Polyglycolide lactide (PLGA) and polyglycolic acid (PGA) are used to make tendon scaffold; PLGA is used in external sheath, while PGA is used in internal core to keep the scaffold straight [3, 7].

The weft-knitted artificial hybrid skin offers more advantages in tissue regeneration with enhanced mechanical properties and reduced wound contraction. The hybrid scaffold (Fig. 8) for tissue regeneration is made with hybridization of knitted PLGA mesh with bovine collagen or chitosan/collagen [7].

Radiotherapy stent is a very special composite structure which is a combination of woven and weft-knitted structures. First, the woven catheter is made on a braiding machine, and after coating and heat setting, the special weft knitting machine is used to knit the weft-knitted warp/weft-inlaid structures (Fig. 9). Collagen, polylactic acid (PLA), polyglycolic acid (PGA) and polyglycolide lactide (PGLA) materials are used to make radiotherapy stent. Multilayered biaxial weft-knitted (MBWK) fabrics which are made by inlaying warp/weft yarn in weft rib structures can be used to make orthopedic splints. Glass yarn or other high-performance yarns are used for inlaying in multilayered biaxial weft-knitted (MBWK) fabrics, while main weft knit structures are made of Polyethylene terephthaalate (PET).

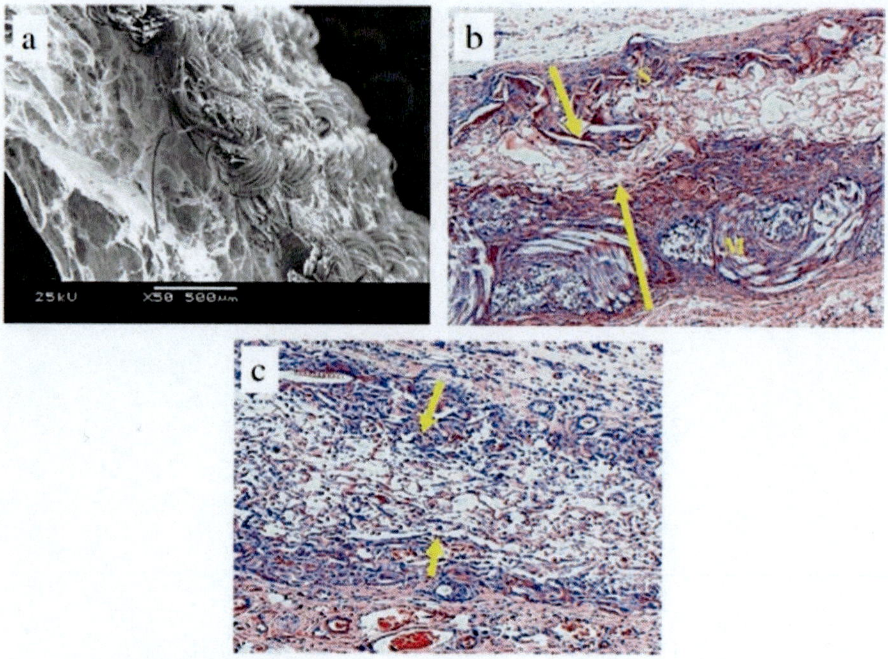

Fig. 8 (**a**) Hybrid PLGA-knitted chitosan/collagen mesh for tissue regeneration (**b**) CCS (**c**) after 1-week implantation [7]

1.3 Healthcare and Hygiene Products

The thermostatic textiles in surgery, medical mattresses, mattress covering materials and protective products fall under the category of healthcare and hygiene products.

The thermostatic textiles in surgery are usually knitted spacer fabrics which are made with warp knitting technology. The purpose of these textiles is to maintain the body temperature of patient during surgery. Special metallic yarn is used in spacer fabrics to generate heat with the help of batteries.

Warp-knitted spacer fabrics are also used to make medical mattresses (Fig. 10); the PET yarn is used to make medical mattresses. Mattress covering materials can be made using both warp and weft knitting structures. The weft-knitted coverings are made on circular weft knitting machines using spacer structures, and the materials used are PET and PA.

2 Sports Applications

The sports goods are used worldwide not only in the field but also outside the field with changing trends. The sportswear has certain qualities that have made it more popular in daily life.

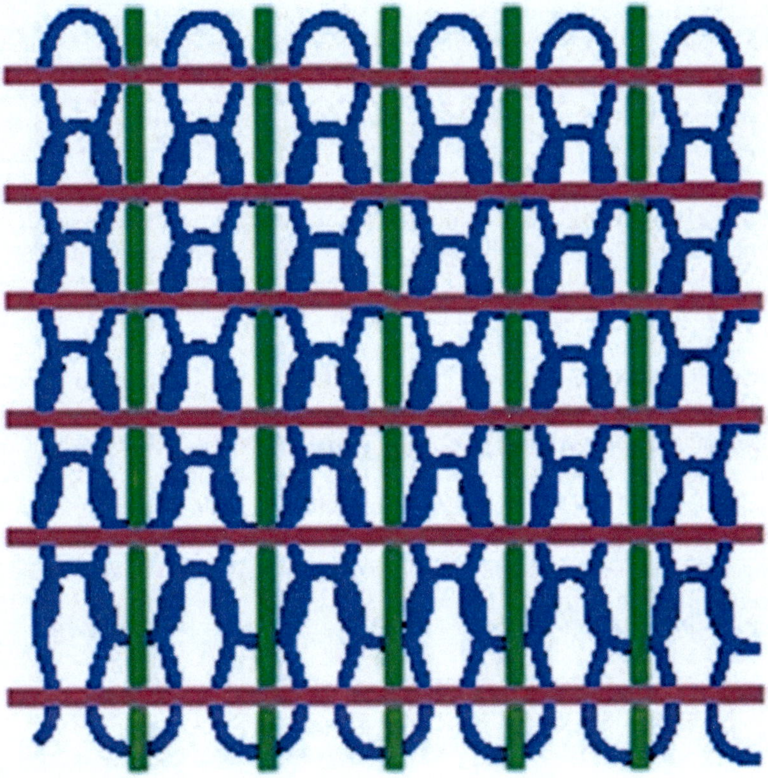

Fig. 9 Weft-knit inlaid fabric used in radiotherapy stent [3]

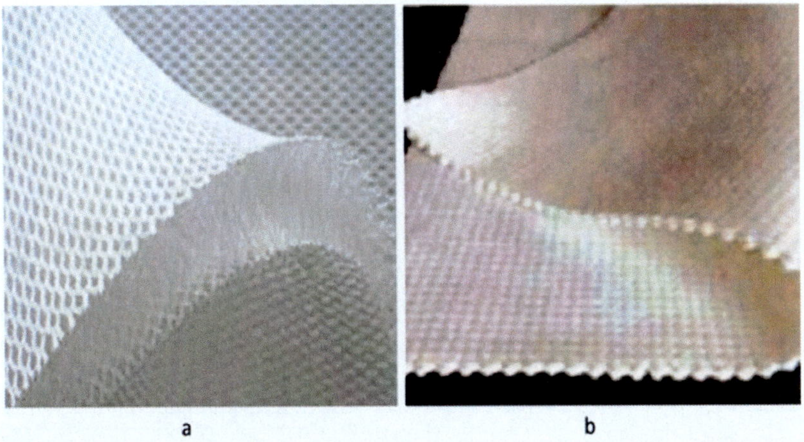

Fig. 10 (**a**) Warp-knitted mattress (**b**) mattress covering

1. These are adjustable and do not restrict body movement.
2. The zoning and mapping of different parts of garments are according to the physical activity and usage of the part, e.g., high sweating areas.

These features can be gathered in single garments by using knitted fabrics. Knitted fabrics are more versatile and easier to use. The reason behind the flexibility of knitted fabrics is the intermeshing of loops which restrict the yarn to only a few places. The woven structures unlike the knitted structures are interlaced, so the movement of yarn is rather difficult, hence less comfortable and breathable.

The techniques which are used to make knitted sports garments are cut and sewn, stitch shaped and cut, fully fashioned and integral. Figure 11 shows the general layout followed to make sports garments from knitted fabrics. The body size machines are used to make tubular fabrics of various sizes. The number of working needles, diameters and loop size determines the size of knitted tubular fabrics. The produced fabric is converted into garments with no side and back seams, but sleeves and gusset are attached to it. This does not hold true for whole garment knitted fabric [8]. The integral knitting technique is used in knitted sports garment manufacturing these days with little to no seam and are made as one piece. The fully fashioned knitting technique is not used mostly for sports garments as the garments made with this technique are usually heavier and not economical.

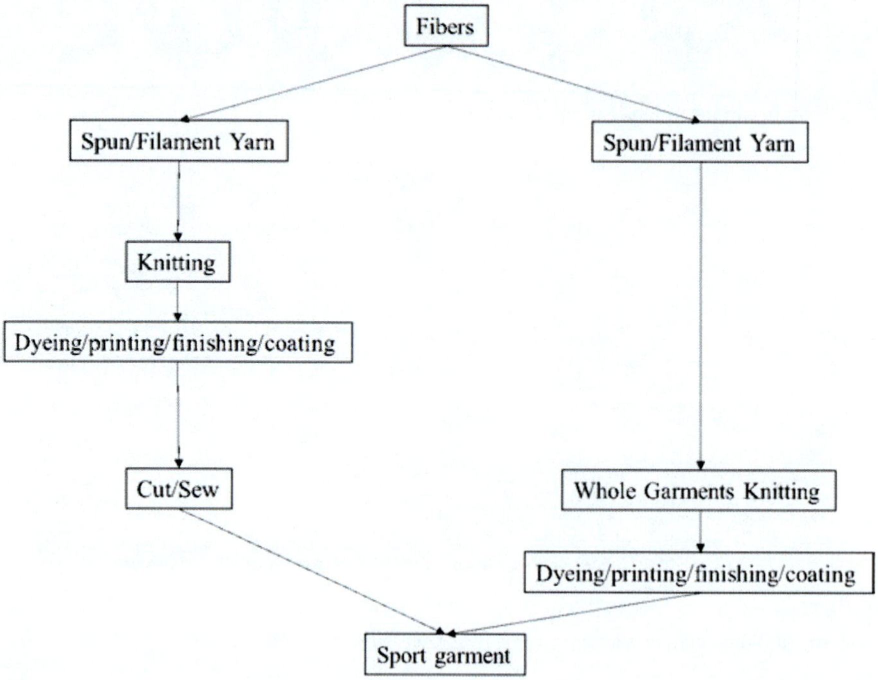

Fig. 11 Flowchart of knitted sports garment manufacturing

Comfort is the most sought-after aspect in modern-day sports garments. Thermo-physiological comfort is one of the important aspects of sports garments especially in performance sports such as swimming, cycling, athletics, bodybuilding, etc. The heat produced during metabolism and gained from the environment must have a balance between them. This is achieved by thermal insulation. The structure of knitted garments has the ability to trap air and provide better thermal insulation. Further, using this ability of knitted structures, thermal insulation can be enhanced by using different types of fibers, structures and densities.

Another aspect of sports garments is mechanical sensation, which is technically called tactile or neurophysiological comfort. It is determined by the type of material used to make garments, type of physical activity, wearing conditions and garment adjustment. Knitted sports garments can deal with this problem. Adding different patches in single garments depends upon the nature and physical activities of different body parts, e.g., elbow, knee, back, etc. as shown in Fig. 12; in particular, in

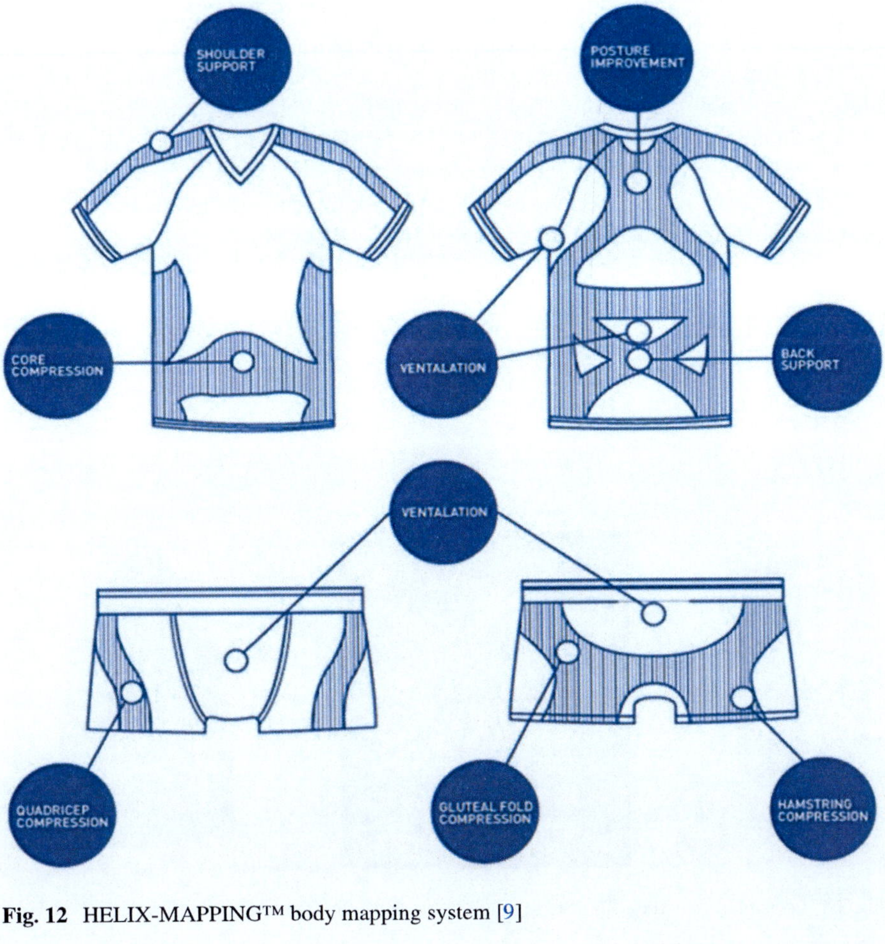

Fig. 12 HELIX-MAPPING™ body mapping system [9]

seamless garments, this is possible within single garments without introducing seam. This has also increased the comfort of garments. The performance of athletes is also increased with the scientific manufacturing of sports garments.

The knitted spacer fabric has very good moisture and heat management because different layers offer different attributes, so spacer fabrics are used in sports shoes extensively. The inner layer which is closest to the skin is hydrophobic, the middle layer has better diffusion and the outer layer is hydrophilic [10].

Different types of machines are used to make knitted sports fabrics, e.g., flatbed, circular and warp knitting machines. Different manufacturers such as Santoni™ (Italy), Shima Seiki (Japan), Stoll (Germany), Orizio (Italy), Fukuhara (Japan) and Mayer & Cie (Germany) have introduced a wide range of machines for whole garments and body size fabric knitting especially used for sports garment manufacturing.

Brands such as Nike, PUMA, Underarmour, Dick's Sporting Goods (DKS) and Adidas imparting many innovations in knitted sports garments keeping in view performance and functionality. Alpha Skin Tech Short Climachill tights are specially designed for hot weather during training and yoga sessions as they keep the body cool and dry (https://www.adidas.co.uk/alphaskin-tech-short-climachill-tights/CF7195.html). Nike Zoom Flyknit shoes are specially designed for sprinters with highly breathable and comfortable tops (https://www.nike.com/gb/t/zoom-fly-flyknit-running-shoe-Dt6Hc9/AR4561-001). Freelite jacket is 100% knitted ski jacket with waterproof and breathable characteristics (Fig. 13).

Knitted garments are used in several sports applications (https://www.kjus.com/fr/en/men-s-freelite-jacket?number=MS15-E07_21000_46).

Fig. 13 (**a**) Alpha Skin Tech Short Climachill tights (**b**) Flyknit shoes (**c**) Freelite jacket

3 Automobiles

Textiles are extensively used in automotive applications such as seat coverings, doors, linings, floor, dashboard, belts, etc. The main requirements of automotive textiles are aesthetics, comfort and safety. Around 5–8 kg^2 (15–18 m^2) of textiles are used in car interior. Both warp- and weft-knitted fabrics are used in automotive applications with a major share of weft-knitted structures. For weft-knitted fabrics, the circular knitted velour is used for door and seat covers, while flat circular knits are used for rear covers of seats and headliners. The different types of warp-knitted fabrics for automotive applications are tricot (structured or brushed), tricot pile sinker fabrics and raschel-knit spacer fabrics. The raschel-knit spacer fabrics are used for car seat coverings [11].

In automobiles, the applications of knitted structures are in two ways: the first is for non-load-bearing applications and the second is for load-bearing applications in composites. Mostly conventional knitted structures are used in non-load-bearing applications such as interior design. After certain modifications, the knitted structures can be used for load-bearing applications such as wheel hose. The weft-knitted near-net-shaped structures, discussed in detail in Section "Composites", are made by the insertion of straight yarn in different directions and are used for automobile applications. In knitted laid-in structures, the high-performance yarn inserted in any direction experiences less mechanical contact, unlike weaving and braiding. Brittle materials are easier to use with less abrasion and load with laid-in structures.

These days, knitted spacer fabrics (Fig. 14), both warp- and weft-knitted fabrics, are used in technical applications, e.g., automobile, composites, medical, etc., but the properties of structures can be tailored according to the end use by changing the material and structure of spacer fabric [11].

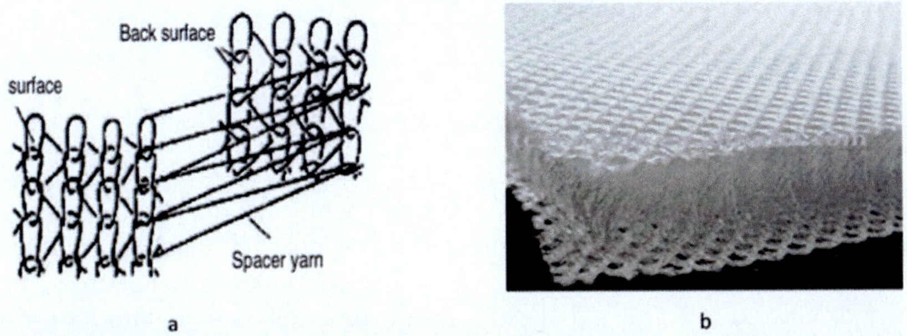

Fig. 14 (**a**) Spacer fabric structure (**b**) spacer fabric sample [12]

3.1 Acoustic Automotive Textiles

The acoustic range of noise within automotive is in the range of 100–4000 Hz, predominantly between 100 and 600 Hz. This noise causes fatigue while traveling both for passengers and for drivers. This high noise can be dumbed/reduced using knitted structures with added features of aesthetics and comfortability. Both weft- and warp-knitted structures have the potential to be used as acoustic textiles.

3.2 Car Seats

Warp-knitted 3D spacer fabrics are a very good replacement for conventional polyurethane (PU)-coated seat covers which cause immense discomfort for drivers. Recycling of PU-coated seat covers is another big problem. The warp-knitted spacer fabrics are a special type of structure in which a spacer yarn is introduced between the front and back surfaces. The thickness of spacer fabric may vary between 2 and 60 mm. The spacer fabrics have better resilience to compression, good breathability, lightweight and better moisture management. To get good compressive properties, the monofilament yarn is used, which can be polyamide, polypropylene and polyester [12].

Car seat covering fabrics are one of the most exposed parts in cars. The general requirements for any car seat fabric are low cost UV resistant, soil resistant and good light fastness properties. Circular knitted pile fabrics made with pile yarn have greater potential to be used in automotive due to better aesthetic feeling and high air permeability [13].

The weft-knit spacer fabric offers more advantage when it comes to cost of production and technicality. The fabric can be made with more ease with few cones on the machine, and patterning on the front side is also possible.

3.3 Other Applications

The wheel hoses of car are load-bearing part which is sometimes made with near-net-shaped weft-knitted structures. The advantage of using knitted fabrics in this part is that the curve shape can be achieved more easily due to the formability of knitted structures [14]. Sun visors sometimes made with loop-raised warp-knitted fabrics are also used. Knitted structures are also used in dashboard and hoses.

4 Composites

Composite is a combination of two or more materials to achieve the properties of all the constituents. The composites are both natural and synthetic. Textile-based composite materials have changed the dimensions of applications. They have added more versatility to composite structures. The properties can be tailored in any direction as per requirement with additional dimensional stability.

4.1 Structure for Composites

Knitted fabric–reinforced composites (Fig. 15) are not often used compared to other textile-reinforced composites due to low fiber volume fraction and looseness of structure which make them unsuitable for load-bearing applications. Knitted fabrics are composed of intermeshed loops, so, near-net-shaped structures can be produced, which has increased their potential in certain composite applications. Knitted fabrics, however, have poor in-plane stiffness/strength due to intermeshed loops. The above problems can be overcome by different techniques.

Both warp- and weft-knitted structures can be used for composite manufacturing. The weft-knitted structures for composite manufacturing are produced on circular or flat knitting machines. In weft-knitted structures, the problem of stability of structure can be overcome by placing inlay yarn in a certain direction in which strength is required [15].

The properties of weft-knit structures can be enhanced in a certain direction by the use of inlaid yarn. The inlaid yarn can be unidirectional or bidirectional (Fig. 16) [16]. The laying-in for weft double-knit structure is done just by placing the yarn between two layers during knitting on the double-cylinder machine. Other than increasing the fabric volume fraction, the laying-in is also done for yarn which is otherwise not possible to knit due to the stiffness of the yarns, but the properties of the yarns are required to meet the requirement of mechanical performance with formability of knitted structures in certain applications. Tucking of the laid-in yarn

Fig. 15 Knitted fabric–reinforced composite illustration

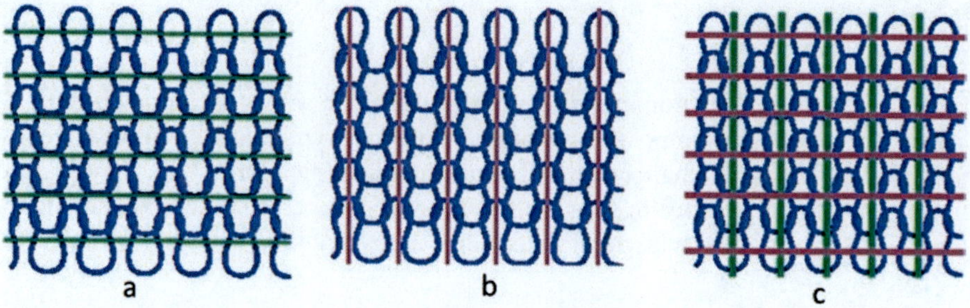

Fig. 16 Knit laid-in structures (**a**) weft laid-in (**b**) warp laid-in (**c**) biaxial warp–weft laid-in structures

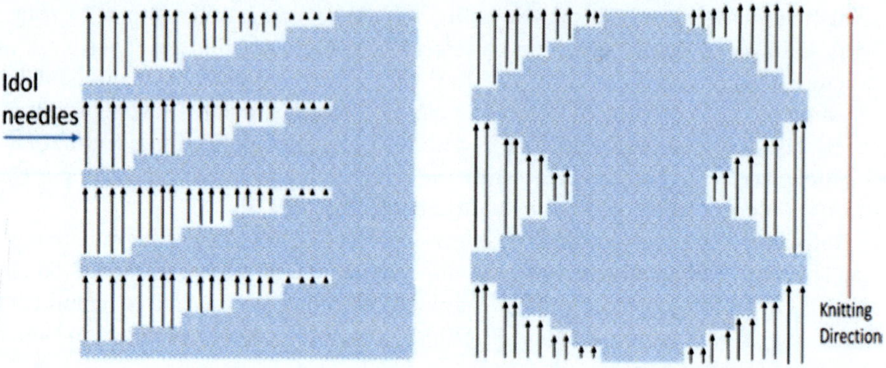

Fig. 17 Shaped knitting on flat knitting machine illustration

can increase the production speed as well as the mechanical performance of structures [14].

Different techniques can be adopted to overcome the traditional problems associated with knitted fabrics for composite manufacturing. In particular, for weft-knitted structures, it is possible to control individual needles on fully fashioned automated flatbed knitting machine.

Holding of yarn on some needles can be possible on fully fashioned flatbed knitting machines, so wedge or triangular shapes can be produced within the knitted fabric. The wedge can be controlled by increasing or decreasing the number of needles and stitch length on each course. These kinds of structures are used for curved parts (Fig. 17) [14].

The expansion and contraction in the knitted structure can be done by playing with the needles. The stitch transfer is a technique used to transfer stitch from needle bed to other needle beds on the flat knitting machine with specialized spring-loaded needles. In flatbed knitting machines, there are specialized cams which help to raise the needle above the normal knit height, so the receiving needle can take the stitch. The racking mechanism is used to transfer the stitch to neighboring needles.

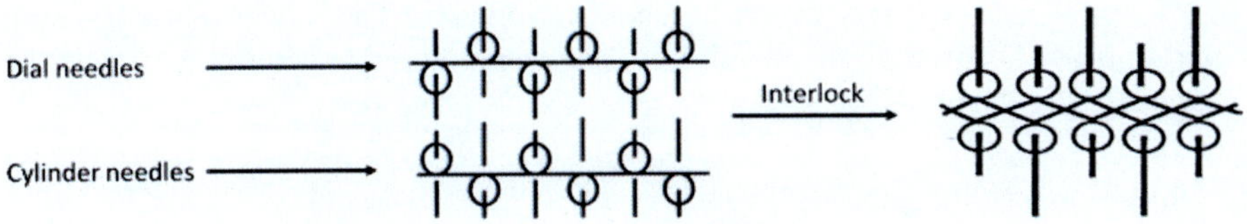

Fig. 18 Graphical representation of interlock fabric stitches on dial and cylinder

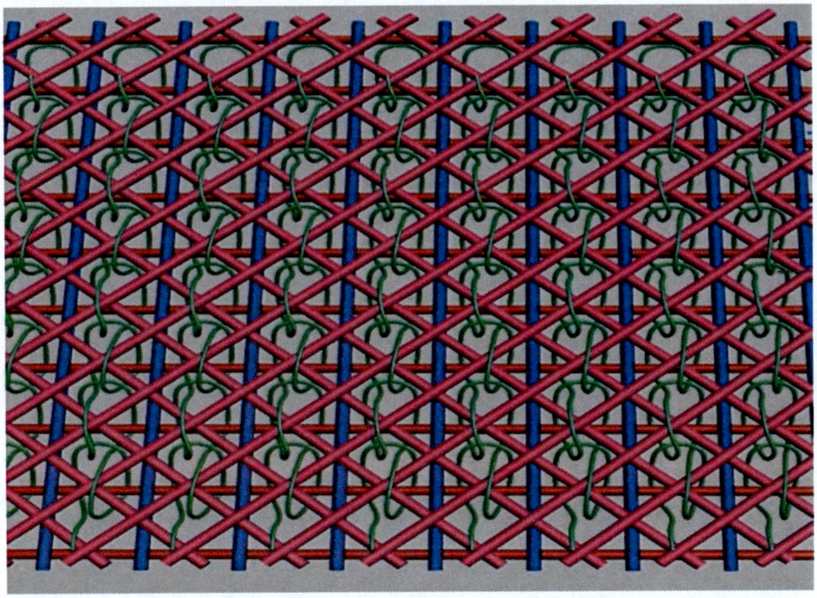

Fig. 19 Warp-knit multiaxial fabric [2]

On the flatbed electronic flat knitting machine, different single-knit and double-knit structures can be introduced after each other in a single length of the fabric, e.g., single jersey and double jersey. Each course can be engineered by selecting and deselecting needles for whole needle bed or partially.

The interlock fabrics (Fig. 18) offer the highest interlaminar shear strength due to their construction. The stitches of two beds (dial and cylinder) are connected evenly.

To cope with the problems of 2D warp- and weft-knitted structures, the multilayer multiaxial warp- and weft-knitted structures were developed with excellent dimensional stability and less extensibility. The multiaxial warp-knitted structures (Fig. 19) are specially developed structures with the advantage of tailoring properties by placing the yarn in any direction as per requirement. These structures are used for different high-tech applications.

The 3D spacer warp- and weft-knitted fabrics are also used for composite manufacturing. In weft-knitted spacer fabric, two layers of fabric are joined together by tuck loops. The tuck stitches work as a connecting yarn. The spacer fabric has high interlaminar shear strength, so sliding between two connected layers can be prevented.

4.2 Problems with Knitted Composites Structures

The knitted fabric composites are not commonly used in industry for composite applications due to low achieveable fiber volume fraction and highly looped structure. The cause of low mechanical properties in composites is stress that is transmitted to only short length as loops are away from each other. Stress is not properly transmitted to fibers, but it is transferred to the matrix resulting in low mechanical properties.

Another reason for lower performance of knitted composite is the breakage of filaments due to excessive bending and tensile force faced by high-performance fiber during knitting on the machine.

4.3 Difference Between Woven and Knitted Composites

There is a long debate about the properties and potential use of knitted fabrics for composite manufacturing. The knitted composites are neglected mainly due to their low mechanical performance, but this does not hold true in all cases though the tensile and compressive properties of knitted fabric–reinforced composites are lower than woven and other textile-reinforced composites; however, knitted fabric composites have good energy absorption characteristics and ease of processing [15]. Table 1 shows the comparison of different properties of woven and knitted composites.

The knitted composites are more prone to crack initiation due to construction of reinforcement, which is at the intermeshing point of loops. When compared, the impact energy absorbed by the knitted composites is higher than that absorbed by woven composite at a similar V_f (50%). The damage area of knitted composite was higher almost six times than woven composite. Another interesting feature which is

Table 1 Comparison of properties of knitted and woven textile composites [16, 17]

Properties	Woven	Knitted
Stiffness and strength	High	Low
Interlaminar fracture toughness	Low	High
Processability of complex shapes	High	Low
Isotropy	Low	High
Material and production cost	High	Low
In-plane properties	High	Low

compression after impact (CAI) strength, decreased 12% for knitted composites as compared to woven composites which is 40% [15].

The interlaminar fracture toughness of knitted fabric composites is higher than woven fabric composites. This high interlaminar fracture toughness is due to looped structures as layers intermingle with each other in such a way that they cannot be further distinguished.

The processing of knitted fabric for composite manufacturing is excellent when it comes to complex shapes with curves and corners. When woven fabric is used in complex shapes, there are often wrinkles in it if used directly; so, they are cut according to the shape of the part, but knitted fabrics can be used as it is due to excellent extensibility, which in some cases is more than 100% [17].

Knitting is a cheaper fabric manufacturing process than weaving, so the cost of composites made with knitted fabric will be quite low. There will also be less cutting waste for knitted fabrics, and processing of knitted fabrics is easier than woven fabrics; thus, knitted fabric composite manufacturing requires less expensive machinery [17].

4.4 Applications of Knitted Composites

There is immense potential for using knitted fabric composites in automotive applications especially for curved parts, e.g., car doors, floor panels and bumpers. Solectria Inc. all composites electrical vehicle has extensive use of knitted fabric composites (near-net-shaped knits). Near-net-shaped knitted fabrics provide high drapeability, good for complex shapers and have good resin flow due to open loop structures [14].

In aerospace industry, net-shaped knitted fabric composites have been used to manufacture rudder-tip fairings of passenger aircraft, helicopter door-track pockets and midsized aircraft radomes [13]. It is also reported that multiaxial warp-knitted spacer fabrics have been evaluated for use in the outer skin, blade and ballistic protection of helicopters [18]. The 3D knitted composites are used in wind turbines, aircraft and vehicles [19]. The 2D and 3D knitted reinforced composites have applications as T-joints and T-shape connectors, cones, pipes and I-beams in structural components. 3D weft or warp-knitted structures have also found applications in jet engine vanes, randomes and rudder-tip fairings [20].

5 Electronics and Energy Applications

Knitted fabrics are extensively used in supercapacitors for energy storage. Conductive yarns are used in the design of such structures which enable effective electrical charge storage. At the same time, their flexible nature is a benefit for

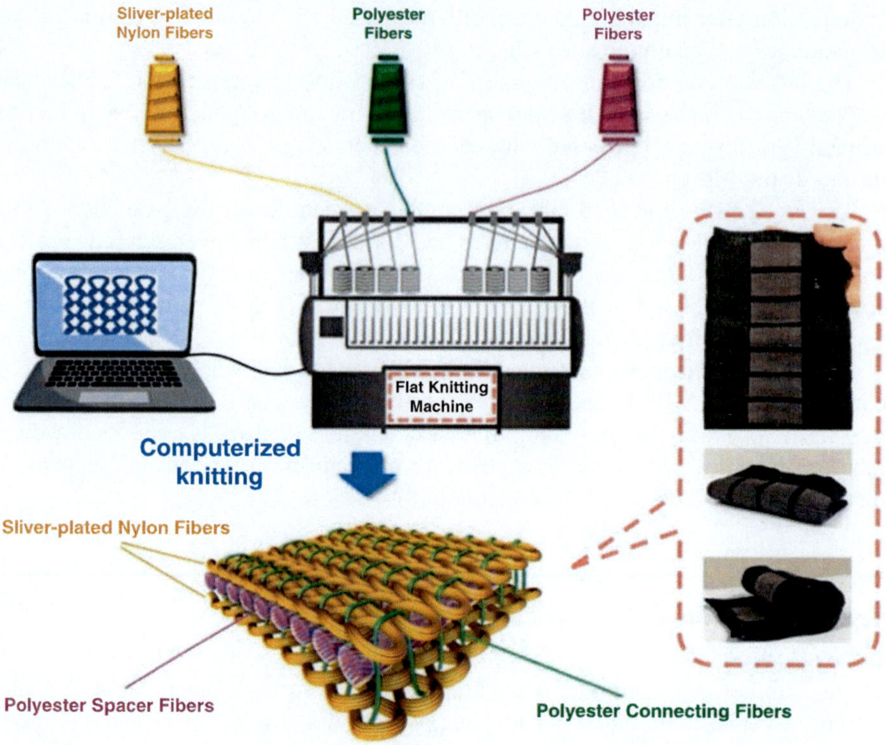

Fig. 20 Knitted fabric–based flexible supercapacitors. (Journal of Power Sources 473 (2020) 228559)

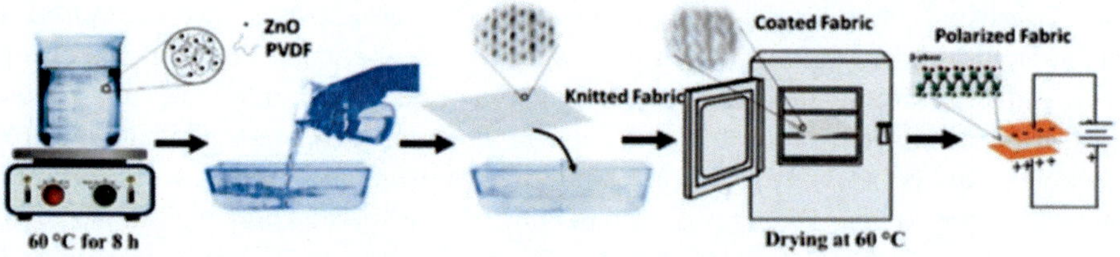

Fig. 21 ZnO–PVDF-coated knitted sensors. (Materials Chemistry and Physics 293 (2023) 126951)

several applications including energy-generating clothing, sportswear, etc. [21]. Spacer-knitted structures used in flexible supercapacitors are shown in Fig. 20.

Smart multifunctional knitted fabrics coated with nanocomposites are used as self-powered sensors. The fabrics are coated with a nanocomposite film of ZnO and poly(vinylidene fluoride) (PVDF) as shown in Fig. 21. Such flexible sensors can produce voltages in the range of 0.6–4.1 V. These flexible sensors are suitable for smart clothing and other electronic devices [22].

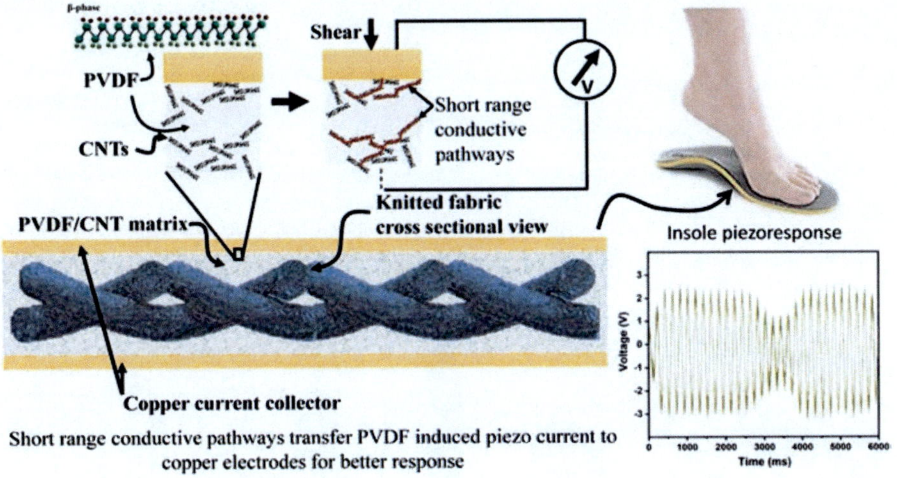

Fig. 22 Flexible sensors based on knitted fabrics for shoe insoles. (Colloids and Surfaces A: Physicochemical and Engineering Aspects 666 (2023) 131279)

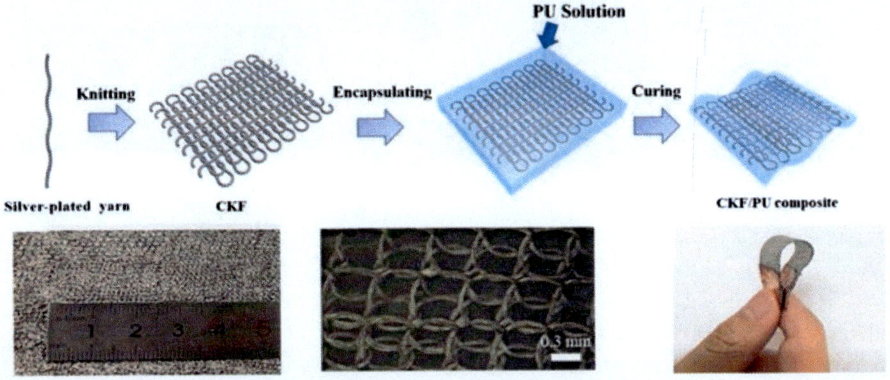

Fig. 23 Flexible knitted fabric–based multifunctional electronic gadgets. (Chemical Engineering Journal 426 (2021) 131931)

Carbon nanotubes (CNTs) are extensively used in PVDF to coat knitted fabrics used in insoles of shoes as shown in Fig. 22 [23]. They enable harvesting of electrical energy during walking and running activities. The knitted fabrics act on the principle of piezoelectric devices. Such flexible textile sensors for energy harvesting show potential applications in medical, diagnostics and other electronic applications.

Knitted fabrics are prepared with conductive coatings for electrical and electronic applications. They are integrated with polyurethane (PU) resin for electronic applications where compression and flexibility are crucial. The schematic and actual sample material are shown in Fig. 23 [24]. These fabrics are widely used in

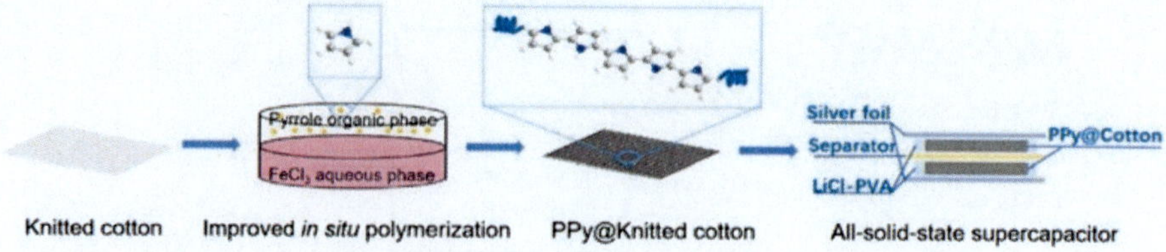

Fig. 24 Chemical deposition of polypyrole nanoparticles on knitted fabrics for electrodes. (Organic Electronics, Volume 74, November 2019, Pages 59–68)

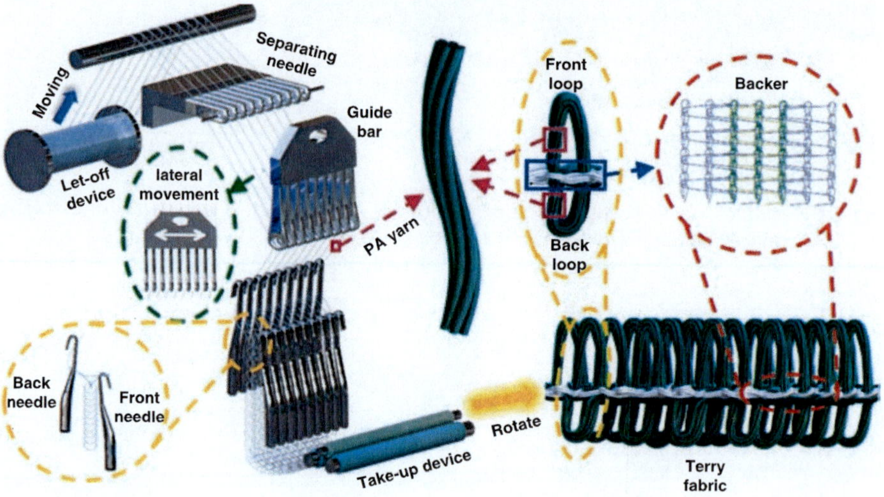

Fig. 25 Fabrication of knitted terry fabric tribological nanogenerator. (Nano Energy, 109 (2023) 108309)

electronic gadgets which are wearable and find suitable applications for military personnel, sports clothing, medical clothing, etc.

Polypyrole-coated knitted fabrics are used as electrodes for energy storage. The lightweight and flexibility of knitted fabrics are an advantage for comfortable wear associated with the use of such wearable electronic systems [25]. The deposition of polypyrole nanoparticles is carried out by chemical deposition technique as shown in Fig. 24.

Knitted terry fabrics are used as triboelectric nanogenerators for motion monitoring and energy harvesting. These structures integrated with the nanogenerators can sense the motion of the body through friction and send signals to the system for further evaluation [26]. The fabrication technique is illustrated in Fig. 25. The nanoscale tribological performance is evaluated by longitudinal and lateral movements of the electrodes.

Similarly, silver-coated conductive yarns are knitted into flexible electronic systems. These are used as self-powered strain sensors. The mechanical deformations in the knitted fabric are sensed by the conductive surface and easily translated into electronic signal. The tribological properties of the fabric are also evaluated by the electronic signals obtained [27]. Due to the 3D arc structure, the sensors are flexible, stretchable and comfortable to wear. The illustration of the fabric is shown in Fig. 26.

Knitted fabrics have also proved to be appropriate as a substrate for telecommunication devices. The conductivity and electromagnetic performances of standard knitted fabrics from cotton and polyester are selected and evaluated as a base material for telecommunication antennas shown in Fig. 27. Flexibility, drape and extensibility are added advantages of such knitted structures in electronic applications [28].

Graphene oxide (GO) and copper (Cu) nanoparticles are applied on nylon knitted fabrics to induce electrical conductivity. Such conductive knitted fabrics are used as motion sensors as well as electrothermal applications [29]. The coating of knitted fabrics with conductive nanoparticles is carried out as shown in Fig. 28. These products have a large potential for medical and electrothermal treatment of skin, muscles and blood vessels.

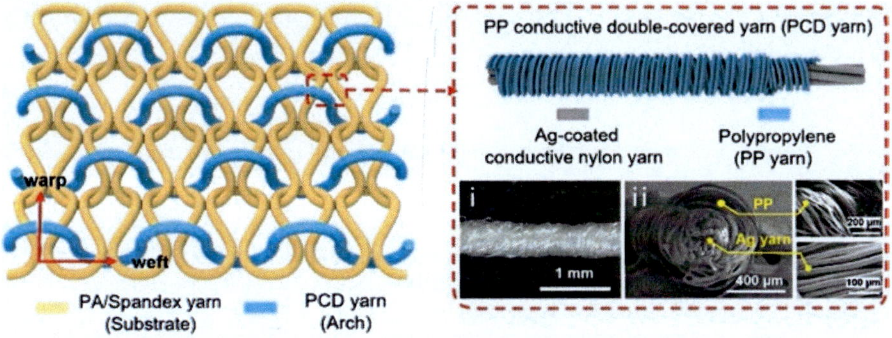

Fig. 26 Knitted fabrics as self-powered strain sensors. (Nano Energy, 109 (2023) 108312)

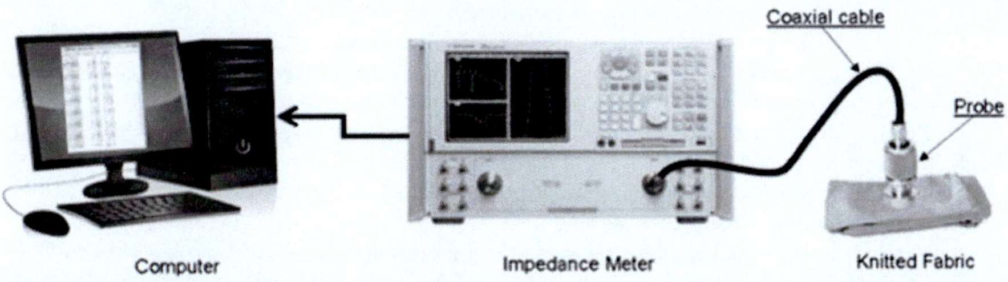

Fig. 27 Knitted fabrics used as substrate for telecommunication antenna. (Journal of Materials Research and Technology, Volume 12, May–June 2021, Pages 2076–2084)

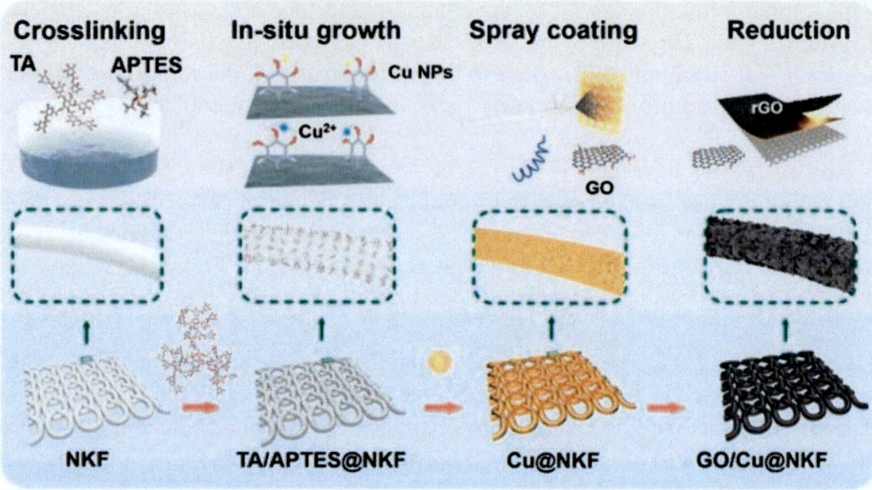

Fig. 28 Treatment of knitted fabrics with nanoparticles for sensing and electrothermal applications. (Surfaces and Interfaces, 36 (2023) 102600)

References

1. V.T. Bartels (ed.), *Handbook of Medical Textiles* (Woodhead Publishing, 2011)
2. M. Hussain, Y. Nawab, Patterning in warp knitting, in *Structural Textile Design*, ed. by Y. Nawab, S.T.A. Hamdani, S. Khubab, (CRC Press, 2017), pp. 231–264
3. X. Zhang, P. Ma, Application of knitting structure textiles in medical areas. AUTEX Res. J. **18**(2), 181–191 (2018)
4. R. Czajka, Development of medical textile market. Fibres Text. East. Eur. **13**(1), 13–15 (2005)
5. R.D. Sumansinghe, M.W. King, New trends in biotextiles – The challenge of tissue engineering. J. Text. Apparel Technol. Manag. **3**(2), 1–13 (2003)
6. *Ultraflex Tracheobronchial Uncovered Stent System – Boston Scientific*. Boston Scientific
7. X. Wang et al., Review article applications of knitted mesh fabrication techniques to scaffolds for tissue engineering and regenerative medicine. J. Mech. Behav. Biomed. Mater. **4**(7), 922–932 (2011)
8. R. Shishoo (ed.), *Textiles for Sportswear* (Woodhead Publishing, 2015)
9. EQUMEN, HELIX-MAPPING™ -Body Response Technology (2018)
10. S. Bruer, G. Smith, N. Powell, Three-dimensionally knit spacer fabrics: A review of production techniques and applications. J. Text. Apparel Technol. Manag. **4**(4), 1–31 (2005)
11. R. Shishoo (ed.), *Textile Advances in the Automotive Industry* (Woodhead Publishing, 2008)
12. P. Taylor et al., Application of warp-knitted spacer fabrics in car seats application of warp-knitted spacer fabrics in car seats. J. Text. Inst. **98**(4), 37–41 (2009)
13. G. B., *Delkumburewatte, Advances in Knitting Technology* (Woodhead Publishing, 2011)
14. A.W. Van Vuurea, F.K. Ko, Net-shape knitting for complex composite preforms AART. Text. Res. J. **73**(1), 1–10 (2003)
15. A.R. Horrocks, S.C. Anand (eds.), *Handbook of Technical Textiles* (Woodhead Publishing, 2016)
16. G. Pamuk, F. Çeken, Manufacturing of weft-knitted fabric reinforced composite materials: A review manufacturing of weft-knitted fabric reinforced composite materials: A review. Mater. Manuf. Process. **23**, 635–640 (2008)

17. B. Gommers, I. Verpoest, P. Van Houtte, Analysis of knitted fabric reinforced composites: Part I. Fibre orientation distribution. Compos. Part A **29A**(98), 1579–1588 (1998)
18. H. Jeon, Ed., *Non-woven Fabrics* (ExLi4EvA, 2016)
19. P. Ma, Z. Gao, A review on the impact tension behaviors of textile structural composites. J. Ind. Text. **44**(4), 572–604 (2015)
20. K. Bilisik, N.S. Karaduman, N.E. Bilisik, 3D Fabrics for technical textile applications, in *Non-woven Fabrics*, ed. by H.-Y. Jeon, (InTech, 2016)
21. J. Wen, X. Bingang, J. Zhou, Towards 3D knitted-fabric derived supercapacitors with full structural and functional integrity of fiber and electroactive materials. J. Power Sources **473**, 228559 (2020)
22. U. Zubair, R. Naseer, M. Ashraf, M. Umer, A. Javid, A. Rashid, Zhebin Xue, multifunctional knit fabrics for self-powered sensing through nanocomposites coatings. Mater. Chem. Phys. **293**, 126951 (2023)
23. U. Zubair, N. Nadeem, H.A. Abid, R.T.M. Ahmad, A. Javid, M. Ashraf, Z.A. Rehan, Highly resilient carbon nanotubes/ poly (vinylidene fluoride) colloidal coated knitted fabrics as proficient sensing and energy harvesting implements. Colloids Surf. A Physicochem. Eng. Asp. **666**, 131279 (2023)
24. G. Yang, R. Xing, Y. Li, C. Ma, B. Cheng, J. Yan, X. Zhuang, Toward high-performance multifunctional electronics: Knitted fabric-based composite with electrically conductive anisotropy and self-healing capacity. Chem. Eng. J. **426**, 131931 (2021)
25. J. Lv, L. Zhang, Y. Zhong, X. Sui, B. Wang, Z. Chen, X. Feng, H. Xu, Z. Mao, High-performance polypyrrole coated knitted cotton fabric electrodes for wearable energy storage. Org. Electron. **74**, 59–68 (2019, November)
26. T. Wang, Y. Shen, L. Chen, K. Wang, L. Niu, G. Liu, H. He, H. Cong, G. Jiang, Q. Zhang, P. Ma, C. Chen, Large-scale production of the 3D warp knitted terry fabric triboelectric nanogenerators for motion monitoring and energy harvesting. Nano Energy **109**, 108309 (2023)
27. X. Fan, X. Jin, C. Lan, Z.H. Guo, R. Zhou, H. Sun, Y. Shao, J. Meng, Y. Liu, X. Pu, 3D arch-structured and machine-knitted triboelectric fabrics as self-powered strain sensors of smart textiles. Nano Energy **109**, 108312 (2023)
28. M. Aquino, J. Guerra, S. Holanda, H. Andrade, M. Melo, L. Zilio, Thiago Santos, Caroliny Santos, an evaluation and characterization study of knitted fabrics for use as telecommunication device substrate. J. Mater. Res. Technol. **12**, 2076–2084 (2021, May–June)
29. Z. Zhao, C. Yan, D. Li, X. Tang, J. Ran, S. Bi, D. Cheng, G. Cai, X. Wang, Fabrication of rGO/ Cu NPs on knitted fabrics for action sensing and electrothermal applications. Surf. Interfaces **36**, 102600 (2023)